Mathmatters

Also by Chris Waring

From 0 to Infinity in 26 Centuries
Maths in Bite-sized Chunks
I Used to Know That Maths
An Equation for Every Occasion

Mathmatters

The Hidden Calculations of Everyday Life

CHRIS WARING

Michael O'Mara Books Limited

First published in Great Britain in 2022
by Michael O'Mara Books Limited
9 Lion Yard
Tremadoc Road
London SW4 7NQ

A CIP catalogue record for this book is available from the
British Library.

Papers used by Michael O'Mara Books Limited are natural, recyclable
products made from wood grown in sustainable forests.
The manufacturing processes conform to the environmental
regulations of the country of origin.

ISBN: 978-1-78929-367-8 in hardback print format
ISBN: 978-1-78929-368-5 in ebook format

1 2 3 4 5 6 7 8 9 10

www.mombooks.com

Cover illustrations by Shutterstock
Designed and typeset by D23
Illustrations by Neil Williams

Printed and bound by CPI Group (UK) Ltd, Croydon, CR0 4YY

CONTENTS

Introduction

Every day, you make thousands of decisions. Some are active, conscious decisions, such as picking up this book and reading these words. Others are instinctive or automatic ones, that you don't realize you are making. These decisions may be formed on the basis of experience, gut instinct, logic, or all three. But logic – and therefore mathematics – underlie all these choices.

The aim of this book is to look at the mathematics of everyday activities and expose the vast world of equations, algorithms, formulae and theorems that underpin them. You can't make a coffee, ride your bike, hire an employee, or even go to sleep without maths being involved.

I'll explain all the maths you'll need to understand as we go along, so don't worry if you haven't thought about all this since you left school. Perhaps you'll find that a little understanding of the mathematics of your everyday life is a powerful thing. It may give you a sense of having more control as well as a sense of wonder at the tiny details which have a huge effect on the outcomes of what you do.

I'll remind you of some basic stuff before we launch into the main event. You don't need to read this part first, but it's here if you need to top up your knowledge.

Ratio

Ratios are used to show proportion, using what maths teachers call 'parts'. For instance, to make purple paint I need to mix five parts red paint with seven parts blue paint. Mathematicians would write this as a ratio – 5:7.

Ratios are useful because they don't rely on any particular quantities in the way recipes do. Whether I'm painting one small wall or the side of a barn, I can use the same ratio.

Surface Area and Volume

Three-dimensional shapes take up space. If you think of a box, there are six faces that enclose the space within it. The area of the six faces is called the surface area and, because the faces are flat, they are measured in squared units: cm^2, m^2 etc. The space within the box is called the volume of the box and, because it is a three-dimensional space, is measured in cubed units: cm^3, m^3, etc. Let's do a quick example:

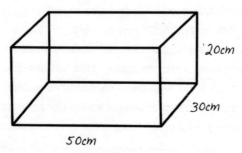

50cm

Here is my box – it is 50 cm long, 30 cm wide and 20 cm tall. Its surface area is the combined area of all the rectangles that make the faces. The area of a rectangle is given by multiplying its length and width. The box has three pairs of rectangles:

$$50 \times 30 = 1,500 \text{ cm}^2$$
$$50 \times 20 = 1,000 \text{ cm}^2$$
$$30 \times 20 = 600 \text{ cm}^2$$

So the total surface area of the box is 2 × (1,500 + 1,000 + 600) = 6,200 cm². The volume of the box is given by multiplying the length, width and height together:

$$\text{Volume} = 50 \times 30 \times 20$$
$$= 30,000 \text{ cm}^2$$

Different shapes require different ways of working out their surface area and volumes, but I'll cover those as needed.

Circles and Spheres

Circles and spheres occur in nature a lot, so it's good to have a handle on how their geometry works. First of all, there is some vocabulary we need to lock down. The distance from the centre to the edge of the circle is called the radius. Twice the radius – all the way across the centre of the circle – is called the diameter.

Long ago, people noticed that dividing a circle's circumference (the distance around the circle) by its diameter always gives the same number no matter the size of the circle. This number is just over 3 – 3.14159265, to eight decimal places anyway – and crops ups in many areas

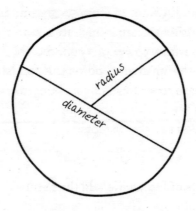

of mathematics. As it goes on forever without repeating, it is represented by the Greek letter pi: π. This letter always reminds me of one of the mighty trilithons of Stonehenge – which itself is designed in concentric circles.

The area of a circle is π multiplied by the radius squared: πr^2.

The circumference of a circle is π multiplied by the diameter of the circle: πd. As the diameter is the same as two radiuses, we can also write: $2\pi r$

Spheres have surface areas and volumes:

$$\text{Surface area} = 4\pi r^2$$
$$\text{Volume of a sphere} = \frac{4\pi r^3}{3}$$

Power and Roots

We've seen several examples of powers already. Powers are the superscript numbers to the right: cm^2, r^3 for example. These numbers are just a shorthand for repeated multiplication. If I want to write 5 × 5 × 5 more concisely, I can write 5^3. We call numbers with a power of 2 *squared* and numbers with a power of 3 *cubed*.

Roots are the inverse of powers. If five squared is 25, then the square root of 25 is five, taking us back to where we started. If $5^3 = 125$, then the cube root of 125 is 5. We use a symbol called a radix for roots:

$$\sqrt{100} = 10$$

For cube roots or higher, we add the number to the radix:

$$\sqrt[3]{8} = 2$$

Pythagoras' Theorem

Right-angled triangles have a special relationship between the lengths of their sides. The longest side of such a triangle is called the hypotenuse and it is always opposite the right angle:

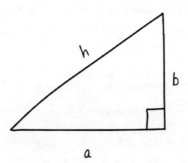

Pythagoras' theorem allows us to calculate an unknown side length if we know the other two:

$$h = \sqrt{a^2 + b^2}$$

If the side you wish to find the length of is not the hypotenuse, you can use:

$$a=\sqrt{h^2-b^2}$$

Speed, Distance and Time

There are two contexts that we do speed, distance and time calculations in: when we are not concerned with acceleration and when we are. In the former case, we can wheel out the simple formulas from school maths: Speed = Distance ÷ Time.

If I catch the train from London to Edinburgh, a distance of 640 kilometres, taking six hours, the speed is 640 ÷ 6 = 107 km/h to the nearest km. This is actually the average speed, as we know the train needs to get moving, stop at stations along the way and maybe goes a bit slower uphill, etc.

If I want to use acceleration, it becomes a bit more difficult. If the acceleration is constant, we can use these formulas:

$$v = u + at$$
$$v^2 = u^2 + 2as$$
$$s = ut + \tfrac{1}{2}at^2$$
$$s = \tfrac{1}{2}(u + v)t$$

In these equations, u stands for the speed at the beginning of the situation, v for the speed at the end of it, a for acceleration, s for the distance and t for the time.

Density

We've all heard the old riddle: which is heavier, a tonne of feathers or a tonne of bricks? The first time you heard it, you may have said bricks. Clearly, a tonne of feathers and a tonne of bricks have the same mass, but bricks are significantly denser than feathers. This is to say that the tonne of bricks will be smaller in volume than the tonne of feathers.

Mass, density and volume are linked by this formula:

$$\text{Density} = \text{Mass} \div \text{Volume}$$

It's worth noting the difference – to mathematicians and scientists, at least – between mass and weight. Although we may use the two words interchangeably in everyday conversation, they have subtly different meanings and values. Mass is a measure of the constituent atoms and molecules that make up an object, measured in kilograms and the like. Weight is a force felt by any object with mass due to gravity, measured in newtons. If you were on the Moon, your mass would be the same, but your weight would be reduced as the gravitational pull is less than the Earth's. On Earth:

$$W = mg$$

Where W is the weight force, m is mass and g is acceleration due to gravity, which is about 9.8 m/s^2. As a rule of thumb, multiply the mass in kilos by ten to get the weight in newtons.

Graphs of Equations

They say a picture is worth a thousand words. Perhaps, for mathematicians, this should be a graph is worth a thousand numbers. Equations can be used to show the relationship between two numbers. Let's take y = x + 1. This is pretty simple and it just tells us that whatever number x is, y is one more. I can show this on a graph. We usually use the horizontal axis to show the value of x and the vertical axis to show the value of y. If I pick a value for x – let's say two – I know that y must one more, which is three. I can plot this point on the graph where x is two and y is three:

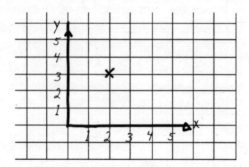

Mathematicians give points coordinates. On the graph above we have marked the point (2,3). I can mark other points where the equation holds true. If I plot all the points, they build up into a line:

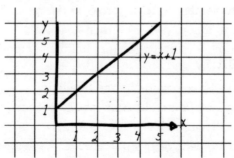

This is useful as I can quickly see the many points where the equation is true. I can extend the axes into negative territory and I can plot more than one line on a graph, and they won't necessarily be straight lines. For instance, here I've added the line $y = 2^x - 5$:

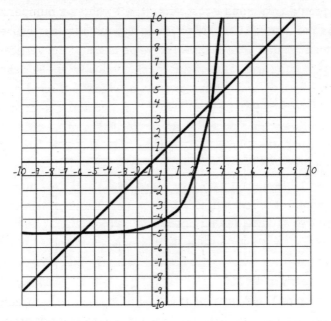

The points where the lines cross show us where both equations are true at the same time, which is useful for solving problems.

Probability

Probability gives a value to the likelihood of something happening. We generally talk about probabilities as a fraction, a decimal from zero (cannot happen) to one (certain to happen), or as a percentage. Some probabilities we can work out mathematically. For these situations we can form a fraction – the probability of an event happening is the number of ways it can happen divided by the total number of possible outcomes. For instance, if I try to roll an odd number on an ordinary six-sided dice, there are three odd numbers I could roll (1, 3, 5) and six outcomes in total (1, 2, 3, 4, 5, 6). This means the probability is $3 \div 6$ which I could write as 0.5, ½ or 50 per cent.

Inequalities

Sometimes, in maths, we know the value of something exactly. For instance, if I said I'm thinking of a number and that half my number was seven, you can work out that my number is 14. However, if I said half my number was more than seven, you can't pinpoint my number but you can say that my number must be more than 14. If we called my mystery number n, then you could write for the first instance that $n = 14$. For the second, you can use an inequality symbol: $n > 14$.

It's worth noting that this means my number cannot actually be 14. If I had wanted to include 14 as an option, I can write it like this: $n \geq 14$. The extra line means that the two values can also be equal.

You can use more than one inequality symbol. For instance, if I said my number was greater than 4 but less than or equal to 9, I could write: $4 < n \leq 9$.

Trigonometry

Just as people noticed the circumference divided by the diameter of a circle always gave the same number, they also recognized similar fixed relationships of measurements in right-angled triangles. They spotted that triangles with the same angles always give the same numbers when corresponding sides were divided. We used to keep tables of these values, which allowed you to work out unknown sides or angles if you had two other pieces of information. Nowadays these tables are programmed into calculators. The three main trigonometric functions we use are known as sine (or sin for short), cosine (cos) and tangent (tan), and θ stands for the angle:

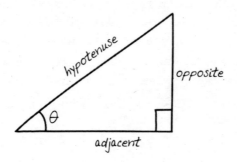

$$\sin \theta = \frac{\text{opposite}}{\text{hypotenuse}}$$

$$\cos \theta = \frac{\text{adjacent}}{\text{hypotenuse}}$$

$$\tan \theta = \frac{\text{opposite}}{\text{adjacent}}$$

You can use these formulas to find a missing side if you know an angle and one other side. If you know two sides and want to know the angle, you need to use the tables backwards, or the inverse trigonometric functions:

$$\theta = \sin^{-1}\left(\frac{\text{opposite}}{\text{hypotenuse}}\right)$$

$$\theta = \cos^{-1}\left(\frac{\text{adjacent}}{\text{hypotenuse}}\right)$$

$$\theta = \tan^{-1}\left(\frac{\text{opposite}}{\text{adjacent}}\right)$$

That's the revision section done with – remember, you can always come back to it if you need to. Now, we're ready to begin our day with a few morning activities and their underlying mathematics.

PART 1

RISE AND SHINE

From the moment you wake up,
mathematics is part of your day.
In the next few chapters we'll take a look
at the governing principles behind your
morning cup of coffee, hitting the gym
and why you sound so damn good
in the shower.

CHAPTER 1

Wake Up and Smell the Coffee

I don't know about you, but I do not consider myself fully functional – or even partially functional, to be honest – until I've had my morning coffee. I silence the spiteful bleating of the alarm and stagger downstairs, barely conscious – barely even a vertebrate. I get out the cafetiere and all the other necessaries while I boil some water. Once that first sip hits my system, the magic begins.

I am not alone. Around 35 per cent of the world's population have a coffee every day, with Scandinavians drinking the most coffee per head. Sixty per cent of adults in the US drink coffee. Four billion cups of the miracle fluid are drunk every year worldwide and the industry is worth over £100 billion. Coffee provides work for over 100 million people, from farmers growing beans to baristas brewing shots.

Coffee became a major thing in North Africa and the Arabian Peninsula in the late fifteenth century, when people invented a drink made from the beans of a bush native to Ethiopia that the local goats and birds loved to chew. The wonderful properties of the drink meant its popularity spread, arriving in Europe in the 1600s with emigrants from the Middle East setting up coffee houses in many cities. From here, cafe culture caught on and coffee trees were planted all over the world.

Previously, Europeans tended to drink alcoholic beverages. The brewing process made the water free of unpleasant diseases such as typhoid and cholera, but the alcohol had serious intellect-dampening side effects. Coffee, on the other hand, improves memory and focus, and so coffee houses became meeting places for intellectuals of every type, independent of class and wealth. Scientists, economists, politicians and revolutionaries would all hold forth in the coffee houses and anyone could join in. Some historians believe that the Enlightenment – the name given to a blossoming of intellectual movements of the seventeenth and eighteenth centuries – started in these coffee houses.

How Much?

The mathematics of coffee brewing is extremely complicated. Modelling thick, lumpy fluids such as coffee – and, indeed, blood and soup – and how they move around is an ongoing mathematical endeavour. We can look at simpler concepts though, such as how much coffee should I put in the cafetiere in the first place? Given my medium-strength ground coffee, I can alter the strength of my drink by altering the ratio of coffee to water. Typical ratios, by mass, vary from 1:10 for a mind-bendingly strong cup of coffee to 1:16 for a paler, less palpitation-inducing brew. I would need to drink more of the weaker coffee to get the same amount of caffeine, and obviously the less water I use, the more intense the flavour of the coffee will be.

How do ratios work? I have discovered through experimentation (I'm nerdy like that) that one gram of coffee for every thirteen of water works well for me. I can show this with the ratio 1:13. Ratios are good because they don't rely on any particular units, provided you take the same type of measurement (for example, mass with mass, volume with volume, etc). The ratio will produce the same strength coffee whether I use one gram of coffee grounds and thirteen grams of water or one ounce of coffee grounds and thirteen ounces of water. Provided that I use the same unit for both the coffee and the water, I'll get the strength of coffee I like, whether I measure it in tonnes, elephants or Chinese *liǎng* (which is equivalent to 50 g, in case you didn't know).

I don't want to be fiddling around with a gram of coffee at a time while I'm standing in my pyjamas on a cold winter's morning, though. Another beautiful thing about ratios is that they are easy to scale up. When I multiply each side of the ratio by the same amount, I'll get different numbers but the same relative proportions. For instance, if I multiply each side of my ideal 1:13 ratio by 20, the ratio becomes 20:260. This tells me I should weigh out 20 g of coffee grounds and pour on 260 g of water. We tend to talk about amounts of water in terms of volume rather than mass – the size of the water rather than how heavy it is – but water conveniently has a volume of 1 millilitre for every gram, so

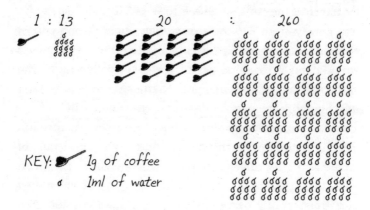

KEY: 1g of coffee
    1ml of water

I need to put 260 ml of water in the kettle. This means I'll get about 260 ml of coffee when I pour, which is enough for a good cup of coffee with room for a little milk.

It's still a bit of a faff to weigh out 20 g of coffee every morning, especially when I can barely keep my eyes open. It's much easier to scoop spoonfuls of coffee grounds into the coffee pot. A heaped dessertspoonful of my coffee grounds weighs about 12 g. How much water does this require? Well, we can see from the original ratio of 1:13 that I need thirteen times as much water as coffee grounds. So, 12 g of coffee will require 12 × 13 = 156 g of water, giving a ratio of 12:156. This clearly won't give as much coffee as the 20 g. If I use two spoonfuls, I double the amounts, giving 24:312. This gives me coffee of the required strength and enough for a bit of a top-up. Great – I can jumpstart my day without the bother of getting the scales out every morning. Of course, I still have to be careful to boil and pour out the right amount of water, but using the right size of cafetiere for your tastes helps with judging this.

Upping Your Coffee Game

I'm sure you know someone who takes their coffee very seriously. Maybe a little too seriously. They may have an espresso machine, and use artisan-ground coffee that's been through a civet's colon; that kind of thing. Mathematically, if I became more serious about coffee, I could introduce other variables to my simple coffee ratio. I could buy beans instead of ground coffee and vary the fineness of the coffee grounds. I could also alter the length of time that the coffee brews for, or change the temperature of the water.

Coffee is chemically very complex and contains over 1,800 ingredients. As the hot water interacts with the coffee grounds, some of those ingredients dissolve faster than others. If you don't brew the coffee for long enough, you end up with more of the fast-dissolving sour flavours, as well as a weaker brew. Leave it to brew for too long and your drink can be overwhelmed by the slow-dissolving bitter-tasting chemicals. Somewhere in the middle is the Goldilocks zone, where sour and bitter combine to make the smooth, rich, caramel flavour that is the hallmark of a good coffee.

The size of the grind affects this, too. The smaller the grind, the faster the coffee will brew compared to a larger grind. Mathematicians at the University of Limerick have been researching the science of making a consistently great cup of coffee. They have modelled the entire process as a system of equations, taking into account the type of roast and the chemistry of the water used, as well as the method of brewing. This complicated mathematical model allowed them to test different combinations without having to physically brew the coffee.

They also made an important discovery to do with the grind size. Many cafes grind their coffee very finely. This works well in espresso machines, where very hot (but not quite boiling) water is forced through the coffee grounds at high pressure to make a very intense 'shot' of coffee. The model showed, however, that if the coffee is too finely ground, the granules clump together and behave like much larger granules, which reduces how much coffee can dissolve after all. The solution is, therefore, to use a slightly larger grind size, which also reduces the amount of coffee required. Customers get a better cup of coffee, cafes use fewer coffee beans, and the environmental impact of the coffee industry is reduced.

Breakfast of Champions

Coffee makes you feel like a superhero, but how much coffee would you need to fuel actual superhero powers?

Even without dipping into the pages of a comic book, caffeine is a stimulant that acts on your central nervous system to produce several effects. The stimulation puts your body into fight or flight mode, so adrenaline is released into your bloodstream. This hormone increases your heart rate and blood pressure, expands the air spaces in your lungs, dilates your pupils to let in more light and diverts blood to your major muscle groups.

Exactly like a superhero.

There's more, though. The caffeine also interferes with your brain's interaction with another hormone called adenosine, which is crucial for alertness. Adenosine accumulates in your brain while you are awake. The more adenosine detected by your brain, the sleepier you feel.

Caffeine stops your brain from detecting the adenosine, making you feel more awake, alert and ready to do tricky maths problems.

So, coffee will make me super-fast and super-clever, with super reactions and super strength, but what if I want some proper superpowers, like being able to shoot lasers out of my eyes?

A normal torch emits light of different wavelengths (which we interpret as colour) that fan out from the torch in a wide beam. In a laser, the light is all of the same wavelength and travels in the same direction, giving a nice tight spot that will drive your cat – literally and figuratively – up the wall. Light is a form of energy, so a powerful laser can be used to burn or cut things. Laser eye surgery uses this idea to make very fine cuts into your eyeballs. But I want lasers to come *out* of my eyes.

Imagine the maths hero QED has to rescue his sidekick Square Root from the evil clutches of their arch-nemesis The Guesser. Square Root is trapped in a steel box suspended over the lava pool in The Guesser's volcanic headquarters. Only QED's laser eye vision can save the day!

Lasers are rated in terms of their power. A laser cutter that can handle steel would typically have a power rating of around 5,000 watts. A watt is a unit of power that represents converting one joule of energy from one form to another every second. So QED's laser vision needs to be capable of converting 5,000 joules of laser power into heat to cut the steel, every second. What's a joule? A joule is a unit of energy that was originally defined in terms of electric charges in circuits, but is now used as the standard unit of energy.

Let's say it will take QED 30 seconds to cut a sidekick-sized hole in the trap. This means $30 \times 5,000 = 150,000$ J will be required, or 150 kJ. How much coffee would this require? Well, one cup of black coffee typically contains about 5 calories. Calories are a unit of energy that we use for food, with one calorie equivalent to 4,184 J. So a

Coffee Machine

Two Hungarian mathematicians, Alfréd Rényi and Paul Erdős, are attributed with the quotation:

Mathematicians are machines for turning coffee into theorems

Both were prolific drinkers of the beverage as well as prolific producers of theorems. So perhaps, if you find mathematics tricky, you should have a cup of coffee before getting started.

5-calorie cup of black coffee contains 5 × 4,184 = 20,920 J of energy. To get 150,000 J I would need 150,000 ÷ 20,920 = 7.17 cups of coffee.

That's quite a bit of coffee and QED doesn't want the jitters when she's in a boss battle. What about if she added some milk? A quick look at the full-fat milk in my fridge tells me it has 276,000 J per 100 ml. This is about ten times as much in the black coffee, in just 100 ml! If she adds 20 ml of milk to her coffee, this increases their energy by 276,000 ÷ 5 = 55,200 J, as 20 ml is one-fifth of 100 ml. This gives each cup 55,200 + 20,920 = 76,120 J. Now she needs 150,000 ÷ 76,120 = 1.97 cups of coffee. Much more manageable.

If QED wants to have some spare laser power for later in the day, she could add sugar to her coffee. Adding two spoonfuls of demerara sugar to each cup would add 60 calories. Converting this to joules give 60 × 4,184 = 251,040 J, bringing her two cups of sweet white coffee up to 2 × 76,120 + 251 040 = 403,280 J, enough for 403,280 ÷ 5000 = 80.6 seconds of laser-eye-blast time. Take that, Guesser!

CHAPTER 2

Power (Ballad) Shower

After some much-needed coffee brings my level of consciousness up to almost-human levels, I head for the shower. As humans, we're drawn to water: we like to be near seas, rivers and lakes, and will pay extra to live by them; we love splashing about in swimming pools, wading in streams and running under sprinklers in parks (although this should probably be limited to toddlers). No one is entirely sure why we love it so much, but theories range from us being descended from swimming apes to water reminding us of the safety and warmth of the womb.

Weird Water

When most substances freeze or solidify, they contract, reducing in volume. Not water. It increases in volume when it freezes, which is what ruptures pipes in cold places. As ice takes up more space than the equivalent mass of water, it is less dense and so it floats in water. This is a big deal, and not just for cooling your gin and tonic. We know that life on Earth started in the oceans, and also that the planet has undergone various Ice Ages. Ice floating on water allows life to survive in the liquid water underneath. If ice sank, it would destroy the life on the sea floor and the rest of the sea would freeze, too. Floating ice actually insulates the water below it, enabling the cold waters of the world to support life year-round.

Another way that water is weird is in its unusually strong surface tension, which refers to the forces in the 'skin' of a liquid. Within the body of a liquid, molecules attract each other in three dimensions due to differences in electrical charges, known as polarity, which means they are pulled equally in all directions. At the surface of the liquid, the molecules are missing the pull from above. This results in them being pulled into the liquid by the molecules below them. This inward pull creates a force on the surface which varies according to the composition of the liquid. Water is second only to mercury – that equally weird liquid metal – for its surface tension. The tension is so strong that small things that are actually denser than water can be made to float if they are carefully placed on the surface. Think of that experiment you did at school with paperclips and how bugs such as water boatmen and pond skaters can stand and glide across the water's surface.

Spheres for the Win

Water forms spherical droplets because of surface tension. As we saw before, the force at the surface of a liquid pulls inwards and so makes the surface area of the water as small as possible. Which shape has the smallest surface area for a given volume? To start with, let's look at a 1 cm cube, as its volume and surface area are easy to work out.

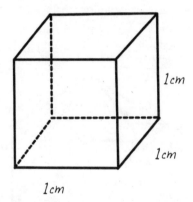

This cube is 1 centimetre on an edge, giving it a volume of 1 cm³. The surface area of the cube is the sum of the areas of all its faces. Each face is a square with an area of 1 cm². Six faces mean a total surface area of 6 cm².

Let's compare this to something with fewer sides. A triangular-based pyramid – or tetrahedron, to use its gangsta-rap name – has four triangular sides:

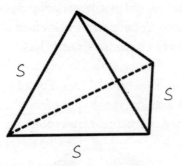

To be 1 cm³ in volume, its surface area has to be 7.21 cm² – larger than that of the cube.

This shows that reducing the number of sides has increased the surface area required. This implies that if I increase the number of sides, the surface area required should be reduced. Let's jump ahead from the six-sided cubes to the twenty-sided shape called an icosahedron:

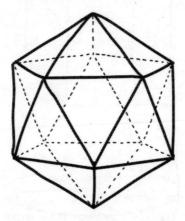

The icosahedron's faces are also made up of equilateral triangles like the tetrahedron. A 1 cm³ icosahedron has a surface area of 5.15 cm². This gives us the smaller result that we were expecting.

While not completely mathematically rigorous, you can see that, for a constant volume, the surface area of a shape seems to decrease the more sides it has, and vice versa. With more than twenty sides, though, we can no longer make shapes with identical faces. For example, here's a 32-sided shape known to mathematicians as a truncated icosahedron or, to normal folk, as a football:

Extended to its limit, as you continue to increase the number of sides, you end up with shapes that look increasing like a sphere:

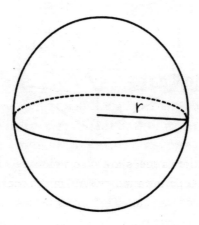

A sphere doesn't have any flat faces as such, but I can still work out its surface area in a similar way. The volume of a sphere is:

$$\text{Volume of a sphere} = \frac{4\pi r^3}{3}$$

In this formula, r is the radius of the sphere – the distance from the centre to the surface. Pi or π you'll remember from school as the number that helps with all things circular and is what you get when you divide a circle's circumference (the distance around the outside of the circle) by its diameter (the distance across the middle of the circle).

If the volume of the sphere is again 1 cm³, a bit of rearranging tells me that:

$$r = \sqrt[3]{\frac{3}{4\pi}}$$

This gives the sphere a radius of 0.62 cm. The surface area of a sphere is given by:

$$\text{Surface area} = 4\pi r^2$$

Substituting 0.62 into this formula gives a surface area of 4.84 cm², smaller even than the icosahedron.

Crocodile Tears

It's a common misconception that falling water takes on a teardrop shape. In fact, the water is usually falling too quickly for us to appreciate it as anything other than a streak, whether from a shower or a cloud. As droplets fall, air resistance actually pushes their spherical shape into the less iconic bun shape:

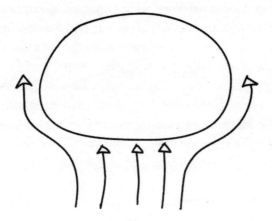

If the droplet is large enough, air resistance will turn the bun into more of a parachute shape and eventually break the droplet, which means that falling water droplets are all less than about 4 mm in diameter. A raindrop is pulled down by gravity, but slowed by air resistance. As the raindrop speeds up, the air resistance increases. When the air resistance matches the gravity, the raindrop has reached its terminal velocity, which is usually about 9 metres per second.

Power Shower

The droplets coming out of my shower, however, are being sprayed, rather than dropping from the clouds. My water-efficient shower uses round 9 litres of water per minute, and I can change the settings of the showerhead to vary the size and number of apertures.

This gets me thinking about how fast the water must come out of the shower. Again, let's simplify things and consider a basic shower with only one aperture for the water to come out of. If 9 litres per minute goes into the showerhead, 9 litres must come out every minute. We can work out the speed of the water from this.

If the aperture was a circle 1 cm in diameter, then 9 litres of water would be forced through this in a minute. Imagine those 9 litres of water in the shape of a 1 cm diameter cylinder. The volume of a cylinder is:

$$\text{Volume of a cylinder} = \pi r^2 l$$

In the formula, r represents the radius of the cylinder and l its length. The volume must be 9 litres, which is 9,000 cm^3, and the radius of the cylinder is 0.5 cm (half the diameter):

$$9{,}000 = \pi \times 0.5^2 \times l$$

Rearranging gives:

$$l = \frac{9{,}000}{\pi \times 0.5^2}$$

This is just less than 11,460 cm or 114.6 m. That is a very long cylinder, but it helps us to find the speed of the water: if 114.6 m of water comes out of the showerhead in one minute, then the water must be travelling at 114.6 metres per minute. This corresponds to 1.9 metres per second – much slower than the 9 metres per second typical of rain. If I want a shower that will ease the knots in my shoulders I've got from slaving over a hot keyboard writing maths books, I'll need to reduce the radius of the aperture. Repeating the calculation for a 1 mm diameter aperture gives a velocity of 190 metres per second, which is about half the speed of sound! Perhaps that's a bit too far the other way.

However I set the shower head, the water doesn't stay at the calculated exit speed for long. The water droplets hit the air, which slows them down, breaks them up and causes them to spread out. Small apertures with high exit velocities will form needle-jets or mist, whereas larger apertures will give slower but wider streams of water.

The water transfers some of its velocity to the air in the shower cubicle. This causes the pressure in the cubicle to drop, compared to outside it. Visible evidence of this: if you have a shower curtain, the pressure difference will suck the curtain into the shower, giving you that clammy touch that no one wants.

Good Vibrations

That cold shock does, however, give me the impetus I need to reach a high note in the song I am belting out. To my ears, with my eyes shut and loofah microphone held aloft, I sound at least as good as Taylor Swift. Doing karaoke after a drink or three, I don't sound nearly as good. Could there be a mathematical reason for this?

The answer lies in the acoustics of a typical shower. Apart from the curtain, most shower spaces are lined with tiles, or other waterproof materials, that reflect sound really well. This means you get a number of echoes as the soundwaves bounce around your cubicle, hitting your ears multiple times from multiple directions.

While I am Shaking It Off, sound is leaving my mouth at about 330 metres per second – the speed of sound. It takes very little time for that sound to reach my ears, especially as some of the sound will travel through my body, where the speed of sound is even faster.

The sound that bounces off the shower wall once, shown on the left above, will reach my ear slightly after that. How long after? Well, I reckon that the wall is half a metre away from my ear, the sound will have to travel about one metre there and back.

Remembering:

$$\text{Speed} = \text{Distance} \div \text{Time}$$

I can rearrange this to get:

$$\text{Time} = \text{Distance} \div \text{Speed}$$

Or $1 \div 330 = 0.003$ seconds or 3 milliseconds.

The sound on the right bounces across the shower several times, travelling about 3 metres in total, so it will take three times longer than the sound on the left – about 9 milliseconds.

The Four Elements

Before we knew about the hundreds of elements in the periodic table, it was believed there were four elements – fire, earth, air and water. The Ancient Greek mathematician Plato believed that these elements corresponded to various three-dimensional shapes now called the Platonic solids, which all have a regular (equal-sided) shape as their face. The painfully spiky tetrahedron represented fire; the solid, stackable cube represented earth; the air consisted of tiny octahedrons (think two square-based pyramids back-to-back) which flow past each other. Water was represented by the twenty-sided icosahedron for its droplet-like shape. The final platonic solid, the twelve-sided dodecahedron, which has pentagonal faces, represented the shape of the universe.

Sound travels in every direction, taking many different length routes to reach my ears. Thus the echoes build together and overlap, causing a phenomenon known as reverberation or reverb, which sound engineers deliberately add to vocal tracks. The overall effect makes your voice sound smoother as the echoes cover up any wobbles in your voice, as well as evening out any pitchy out-of-tune singing. It also gives the illusion that you are singing in a much larger space – belting out your favourite tune in a hip venue to a rapt audience, rather than to your rubber ducky in your mildewy shower.

There is another effect at play, called resonance. Sound

is vibration, and any object that vibrates has frequencies that it is better at producing than others. These resonant frequencies vary according to the geometry and materials of the vibrating thing. Showers, it turns out, typically resonate at human bass vocal ranges. Essentially, showers add a bass boost to your voice which makes it sound fuller and richer, even if you're trying to hit that key change in 'Livin' on a Prayer.'

CHAPTER 3

No Pain, No Gain

Whether you enjoy exercise and sport or see it as a chore you endure to remain healthy, we all undertake varying amounts of physical activity. Sport and fitness are big business, too, for the likes of professional athletes and sports people, or sales of protein shakes and isotonic drinks in the supermarket.

Sports science has also come to the fore, as athletes seek advantages wherever they can find them. Experts on human physiology, the best sport scientists can have highly lucrative careers in the employ of teams or national squads searching for trophies and medals.

Like any science, sports science comes with a slew of formulas and equations that aim to make training more efficient and focused. We'll take a look at some of the more common ones to examine the maths that makes them tick. With the rise of technologies like heart-rate monitors and fitness trackers, there are concepts like heart-rate zones and basal metabolic rates that also require some mathematical explanation.

Body Mass Index

Thanks to sports science, many of us are now familiar with the concept of Body Mass Index, or BMI. BMI gives an indication of whether your weight is appropriate for your height. You calculate your BMI with this formula:

$$BMI = \frac{Mass}{Height^2}$$

You need to use your mass in kilograms and your height in metres. For me, at 1.9 metres tall and weighing 90 kilograms, my calculation is $90 \div 1.9^2$, which gives me a BMI of 24.9.

Great. What does this mean? Well, fortunately for me, a healthy BMI is between 18 and 25, so I'm just inside this range. The term Body Mass Index was coined by American physiologist Ancel Keys (1904–2004), but he was quick to say that it was only suitable for getting an idea of a population's BMI, and was not always appropriate for individuals. It only works on fully grown adults, for example, but even then it doesn't take body composition into account. Ukrainian Oleksei Novikov, who was the World's Strongest Man in 2020, is not hugely tall at 1.85 metres, but he weighs a sturdy 135 kg. This makes his BMI over 39, implying that he is morbidly obese, whereas in fact he is extremely muscular. Muscle is denser than fat, giving very muscular people a high BMI.

So, once you've calculated your BMI, you may decide it

would be a good idea to do a bit more exercise, or watch your calorie intake a bit more carefully.

Dropping the Calories

The UK National Health Service recommends a daily intake of 2,000 calories for women and 2,500 for men. These figures are very broad guidelines, however, and won't be appropriate for everyone. If you work in heavy manual labour, you could require double this intake to maintain your weight.

As well as your lifestyle, researchers discovered that there are other key factors in how many calories a person requires: gender, weight, height and age. The Harris-Benedict equations, first produced in 1919, give a value called the Basal Metabolic Rate or BMR:

$$BMR\ (female) = Mass\ (kg) \times 10 + Height\ (cm) \times 6.35 - Age\ (years) \times 5 - 161$$
$$BMR\ (male) = Mass\ (kg) \times 10 + Height\ (cm) \times 6.35 - Age\ (years) \times 5 + 5$$

Using myself as an example again: $90 \times 10 + 190 \times 6.35 - 43 \times 5 + 5 = 1,897$ calories per day to the nearest calorie. This is how much energy my body needs to keep itself going if I lie in bed all day. Sadly, this is not often an option, so I can calculate my daily total calorie requirement by multiplying this BMR by a number between 1.2 (sedentary lifestyle with no exercise) to over 2 if you are a full-time athlete or labourer. Generously scoring my dog-walking and taking the kids swimming at 1.4 gives me a daily calorie requirement of $1,897 \times 1.4 = 2,656$ calories. This is in the same ball park as the generic 2,500 calories recommended for men (2,000 for women). Whether I actually eat this many calories per day is a different question, of course.

The fact that my BMI is flirting with 25 suggests that I have probably consumed more than I need.

If I wanted to lose weight purely by eating less, eating fewer than 2,656 calories would make that happen. Let's say I had a target of dropping 5 kg – how would I go about this? Well, what I want to happen is to lose 5 kg of fat. Pure fat contains about 9 calories per gram, but body fat is not pure and contains around 7.7 calories per gram, so 5 kg of body fat represents 5,000 × 7.7 = 38,500 calories. If I reduce my calorie intake by 500 calories per day (a typical amount recommended by health professionals), this will take 38,500 ÷ 500 = 77 days. Unfortunately, as you lose weight your body becomes more efficient at conserving calories, so it will probably take a bit longer than this to lose the kilos. About three months of eating 2,156 calories a day should do the job.

Go for the Burn

The other option is to burn more calories by doing more exercise. Generally speaking, the more intense the exercise, the more calories it will burn. Walking, for instance, generally burns about 4 calories per minute. So that 5 kg would require 38,500 ÷ 4 = 9,625 minutes of walking. This is about 160 hours, so I would need to walk for an extra 1 hour and 47 minutes per day to lose the weight in the same time frame as the diet alone.

Running burns more like 13 calories per minute; over three times as much as walking. 38,500 ÷ 13 = 2,962 minutes of running. This is means I'd have to run for 33 minutes per day on average to shift the 5 kg in the same time as the diet.

Go with the Flow

The flow of viscous liquids, such as blood, are governed by the Navier-Stokes equations, an example of which is:

$$\frac{\delta v}{\delta t} + (v \cdot \nabla)v = -\frac{1}{\rho}\nabla p + v\Delta v + f(x,t)$$

If this looks difficult, it's because it is! It introduces the idea of turbulent flow, which is one of the great unsolved mysteries of maths and physics. If you can work out a way to solve the equations, you would receive a US$1 million prize from the Clay Mathematics Institute. Get solving!

Whatever exercise I do, there is a trade-off between how intense the exercise is and how long I do it for. The intensity of the exercise is usually measured by comparing your heart rate while doing it to your theoretical maximum.

You can estimate your maximum heart rate with a simple formula:

Maximum Heart Rate = 220 - Age (years)

This gives me a maximum HR of 220 - 43 = 177 beats per minute – roughly three times per second. This value will clearly not hold true for every 43-year-old person, but it should be something near this. If you were a healthy person used to high-intensity workouts, you could exercise at your maximum capacity for as long as possible using a health tracker to see your highest heart rate but this would be a very unpleasant – if not downright dangerous – process for most people.

Most fitness programmes, then, split workouts into various heart-rate zones using percentages. To lose weight, we need to get to a heart rate that means we are burning calories nicely, but that isn't so high that we can't maintain it for long. For most people, this occurs at around 60 to 70 per cent of the max HR. The calculations for me would be:

$$60 \%: 0.6 \times 177 = 106.2$$
$$70 \%: 0.7 \times 177 = 123.9$$

So, if I keep my heart rate between 106 and 124 beats per minute I should be able to exercise for a long time in this fat-burning zone. If I go harder than this, I will burn more calories, but my body will start to produce lactic acid as I get out of breath. This acid is the stuff that makes higher intensity stuff hurt.

Weight-Watching

At the end of my programme, how has my BMI changed? Now weighing a svelte 85 kg, my BMI is $85 \div 1.9^2 = 23.5$. My BMR is now $85 \times 10 + 190 \times 6.35 - 43 + 5 \times 5 + 5 = 1,836$. With my new exercise regime giving me a multiplier of 1.5, I can eat $1,836 \times 1.5 = 2,754$ calories to maintain my new jeans-comfortable weight.

PART 2

On the Move

Once you've got yourself up, fit and ready to face the day, take a look at the mathematics of getting from A to B – whether that's by bike, car, or hypersonic rocket.

CHAPTER 4

It's Like Riding a Bike

Whether it's a cheap way to commute to work, a way to tire the kids out, or your profession, riding a bike is a lovely (and environmentally friendly) way to get around. With improved provision of cycle lanes and shared paths it can often allow you to avoid traffic jams and eliminate parking fees, too.

The mathematics of cycling is that of circles – wheels, cogs and pedal cranks all translate revolutions into linear speed, moving you towards where you want to go.

Penny for Your Thoughts

Many bicycles come with a selection of gears to help you get up hills, but even if they don't, you can cover a good distance quicker than on foot and without too much effort. Gears add complication, weight and expense though. In the 1870s, no one had yet thought to put gears on a bicycle, so they were designed with pedals attached directly to the front wheel. This meant that every revolution of the pedals gave one revolution of the wheels.

Most people find that spinning the pedals around once per second is a comfortable yet ground-eating pace. The wheels of my road bike have a diameter of 70 cm. From this, I can work out how fast I would travel.

The circumference of my wheel is:

$$\text{Circumference} = \pi \times \text{diameter}$$
$$= \pi \times 0.7$$
$$= 2.20 \text{ m}$$

So, if I rode my bike Victorian-style, with one revolution of the pedals per revolution of the wheels, I would cover 2.2 metres every second. I can convert this to the more familiar kilometres per hour by doing this:

$$2.2 \times 60 = 132 \text{ metres per minute}$$
$$132 \times 60 = 7{,}920 \text{ metres per hour}$$
$$7{,}920 \div 1{,}000 = 7{,}920 \text{ kilometres per hour}$$

This is not terribly fast (roughly 5 mph) – a comfortable walking pace is 5 km/h. It will save us time to notice that, to convert m/s into km/h I had to multiply by 60 twice and then divide by 1,000. I can smoosh this into one operation because:

$$60 \times 60 \div 1{,}000 = 3.6$$

Multiplying m/s by 3.6 will give me km/h. Anyway, if I want to go faster, the easiest option is to spin my legs faster. Professional cyclists typically manage 100 revolutions per minute (rpm) on the pedals, which is $100 \div 60 = 1.67$ revolutions per second:

> ## Big Wheels Keep on Turning
>
> The 1876 Leamington Spa Cycle Show witnessed the birth of a monster. James Starley, a British cycling pioneer, was keen to demonstrate his new method of spoke placement. He developed a Penny Farthing with a 78-inch (just under two metres) wheel which, thanks to cleverly placed treadles, could actually be ridden by his son.

$$2.2 \times 1.67 = 3.674 \text{ m/s}$$
$$3.674 \times 3.6 = 13.23 \text{ km/h}$$

This is an improvement, but the typical cycle commuter probably wants to hit at least 20 km/h, which would require a trouser-mangling 150 rpm. I don't want to pedal any faster, so how can I increase my speed?

The Victorians solved this problem by making the front wheel very much bigger, giving us the distinctive – and scary-looking – Penny Farthing, with a front wheel diameter of about 130 cm. The circumference of this wheel is $\pi \times 1.3 = 4.08$ m, so I would cover nearly twice as much ground for each turn of the pedals. 20 km/h is 5.55 m/s and means I would have to spin the pedals $5.55 \div 4.08 = 1.36$ times per second or 80 times per minute. This is still a bit more than the comfortable 60 times per minute, but at least I'm not in Olympic Time Trial territory.

Fortunately, in the late 1800s, the familiar 'safety' bicycle was invented, so-called because the Victorians wisely felt that being able to reach the ground with your feet while on the bike was safer. Having the pedals turn a toothed cog known as the chainring, attached by a chain to another cog or sprocket on the axle of the rear wheel, allowed you to alter the ratio of your leg rotations to the wheel rotations.

We saw before that my 20 km/h commuting speed would require 150 rpm from the wheel, but now I can use the gears to help me. If I give the chain ring twice as many teeth as the sprocket, then one turn of my legs will spin the back wheel twice, meaning my legs only need to maintain 75 rpm. This gives me a gear ratio of 1:2 – 1 rpm of the pedals gives 2 rpm of the wheels. The Penny Farthing was stuck with 1:1.

Wheely Hard Maths

Nowadays, bikes usually have a system which allows you to select different chain rings and sprockets to find the perfect combination for your style of pedalling, the speed you want to go and the terrain you are travelling over.

Typically, a new road bike will have two chainrings, one with 50 teeth and the other with 34. It will then have a 'cassette' of sprockets on the back wheel ranging from 11 to 28 teeth. Using the 50 and the 11 gives a ratio of nearly 1:5, perfect for cycling really fast downhill. At the other end, 34 and 28 is nearly 1:1, making it perfect for getting up the hill in the first place.

It can be difficult to compare ratios: which is harder to push – 50:22 or 34:15? It's hard to tell at a glance. So cyclists will often talk about gears in terms of how far the bike goes for one revolution of the pedals. We can do this with this formula:

$$\frac{\text{number of chainring teeth}}{\text{number of sprocket teeth}} \times \text{wheel circumference}$$

So 50:22 on a road bike wheel with circumference 2.2 metres gives:

$$\frac{50}{22} \times 2.2 = 5 \text{ m}$$

And 34:15 gives:

$$\frac{34}{15} \times 2.2 = 4.99 \text{ m}$$

Not a lot in it. The larger the number, the harder the pedals are to push.

Breaking the Speed Limit

In September 2018, American cyclist Denise Mueller-Korenek broke the speed record for a bicycle, hitting 296 km/h behind a dragster fitted with a large windbreak. She used a bicycle fitted with 17-inch motorbike wheels with a double set of gears: a 60-tooth chainring going to a 13-tooth sprocket, which was on the same axle as another 60-tooth chainring which led to a 12-tooth cog on the rear wheel's axle. How big was this gear?

One revolution of the pedals would turn the first sprocket several times, which then gets transferred to the second chainring, so we multiply the fractions together. The 17-inch wheel had at least an inch of tyre on each side of the wheel, so let's call it 19 inches in diameter. Finally, we need to convert to metres, and each inch is 0.0258 metres:

$$\frac{60}{13} \times \frac{60}{12} \times \pi \times 19 \times 0.0258 = 35.5 \text{ m}$$

That's a big gear, but you'd need it to go this fast.

Circular Arguments

One of the most remarkable things about bicycles is the fact that they stay upright. They seem to like staying upright more than falling over, if the number of teenagers tapping on their phones while riding along is anything to go by. In fact, tests have shown that, if you give a well-made bicycle a hard shove, it will coast along without falling over, even if you push it sideways as it goes.

There are quite a few factors involved in what keeps a bicycle upright, or rather, what stops them from falling over. Originally, it was thought to be the gyroscopic effect of the spinning wheels. The gyroscopic force, essentially, means it requires more force to tilt an object when it is spinning than when it is not. This was found not to be the case when experimenters made bikes with special wheels with contra-rotating parts that cancelled out the effect.

Another factor is the castor effect. If you've ever pushed a bike anywhere, you'll have noticed that, if you lean the bike to the left, the front wheel will tilt to the left also. This means the bike naturally steers into any turn, which has the effect of restoring the bike to balance, albeit in a new direction. Steering is achieved principally by leaning the bike, and turning the handlebars makes the bike lean.

The global bike industry is valued at around £40 billion, so mathematical modelling of how bikes handle is important to manufacturers. It also has implications for understanding how anything stays upright – including people or robots when they walk. So whether you're racing at the velodrome, cycling to work or helping your kids learn to ride, remember that maths is helping you to keep your balance.

CHAPTER 5

Road Rage

For many people, commuting to work involves driving. It can seem that driving a car anywhere in a busy town or city puts your fate in the lap of the traffic gods, especially when your view is limited to whatever you can see out of the windows of your motor. Will the gods be kind and give you that clear run you need to get to work without the grinding tedium of stop-start traffic, road works and phantom traffic jams? Or will you arrive stressed and angry after what feels like a lifetime spent fighting for space on the road as three lanes merge into one?

The mathematics of traffic flow is a well-established field of study with a host of complex theorems and equations that attempt to predict how traffic will behave in a given set of circumstances. The first work comes from the 1920s, just as cars became more commonplace on roads. Fast forward one hundred years and the flow of people and goods on the road network has a significant role in the nation's economy. Planning and maintaining good traffic flow is serious business.

Pit Stop

One of the many necessaries of driving a car is refuelling. Increasingly, hybrid and electric cars make it possible to refuel at home, but most of us still need to pull into the petrol station to fill up.

A driver's refuelling habits are like a personality test. Are you a crazy maverick who drives with the fuel gauge in the red, only putting in a few litres at a time? Or do you enjoy the safety and security of a full tank, knowing that you could drive off into the sunset and beyond? Would either approach save you money?

Having a full tank makes the car heavier, and the heavier something is, the more force is required to accelerate it. If we assume that the cost of fuelling the car is proportional to its mass, we can work out the mass of the petrol in the car as a percentage of the mass of the car, and see what kind of savings might be involved. For this I need some information, so I turn to the car owner's manual of my own car.

The back of the manual has a treasure trove of exciting data, from engine torque to battery-charging profiles. My car has a mass of 1,500 kg and a fuel-tank capacity of 45 litres. But how heavy is 45 litres of petrol?

Petrol has a density of about 750 kg/m^3 – that is, a metre cubed of petrol weighs 750 kg. These units don't match the litres quoted for the fuel tank, so we have to convert them. A millilitre is equivalent to a cubic centimetre, so if we can work out how many cubic centimetres there are in a metre cubed, we can convert between that and litres.

It is very tempting to say that 100 cm^3 must make 1 m^3 because one metre is 100 centimetres. But a quick sketch shows there must be more than 100 cm^3 in a cubic metre.

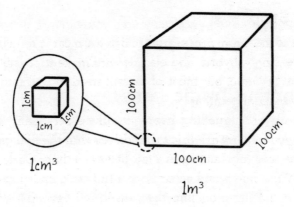

$1cm^3$

$100cm$

$1m^3$

Way more, in fact.

The metre cube is 100 cm on each edge. The volume of a cube is the length of the side, cubed. This means there must be:

$$100^3 = 100 \times 100 \times 100 = 1,000,000$$

Whoa – one m^3 is a million cm^3! A litre is 1,000 ml, which is one-thousandth of a m^3: 1 litre = 0.001 m^3.

Great. So a tank of petrol has a volume of 45 litres, which converts to 0.045 m^3. Each metre cubed of petrol weighs 750 kg, so the petrol in the tank weighs 750 × 0.045 = 33.75 kg. To work this out as a percentage of the mass of the car, we divide it by the mass of the car to get a fraction and then multiply this by 100 to get the percentage:

$$\frac{33.75}{1500} \times 100 = 2.25 \%$$

This is not a huge percentage, even with a full tank. Running with an emptier tank that you top up more often will not save you much money and will increase the number of visits to petrol stations, which may take you out of your way anyway. With a full tank in a small car only weighing around 34 kg, carrying a passenger or three will do way more damage to your fuel economy.

Phantom Traffic Jams

Anyone who has spent any length of time on faster roads will be familiar with this situation: you see a wall of brake lights and join a queue of stop-start traffic, your heart sinking with the speed of the car. Sometimes a lane will gather a bit of speed, but then it will slow to a halt again. You crane your neck to see what the hold-up is – slow traffic at a junction, malfunctioning traffic lights, an accident, or maybe a lane blocked by roadworks or a breakdown. But there isn't anything, and you get to the front of the jam and accelerate away. There was no apparent reason for it, but it happened, nonetheless. Why?

Research shows that these jams are a function of the reaction times of drivers responding to a slower-moving vehicle – often a lorry going up a hill – in their path. Good drivers look ahead, past the vehicle in front of them, ready to react to conditions up the road. They will see the brake lights come on and begin braking themselves shortly thereafter. Some of them may need to brake a little harder that the first driver, but they can still stay a comfortable distance apart and keep the traffic moving. If they do get a bit squashed up, they can cope until the lorry crests the hill or they are able to overtake it. Essentially, the good drivers mitigate the effect of reaction time by reacting en masse to the slow lorry.

Drivers that aren't looking ahead will have to brake harder, as will tailgaters and faster drivers. This causes a chain reaction, with each driver having to brake harder than the one in front of them.

Let's consider a convoy of drivers who only respond to the car directly in front of them. If they were driving along at 25 m/s, they'd need 50 metres between them to maintain a Highway Code-friendly two-second gap. The first car comes upon a lorry only going 20 m/s. This requires the car to slow down and the 2-second gap would be 40 metres. Essentially, they all need to slow down before they gain 10 metres on the lorry. However, the second driver in the convoy needs time to react to the braking of the car in front – about half a second being typical for an alert person. In the half-second reaction time the car in front would get 2.5 metres closer to the lorry, as it is closing at 5 m/s. The second driver needs to brake in a similar fashion to the first, but the 10-metre buffer has now been eroded to 7.5 metres. Rearranging the $v^2 = u^2 + 2as$ formula (see page 12) gives:

$$a = \frac{0^2 - 5^2}{2 \times 7.5}$$

This gives a = -1.67 m/s^2. This is pretty comfortable, but let's keep going for the next few cars. Car 3 needs to decelerate at -2.5 m/s^2, which is fairly hard, being twice the deceleration of the first car, but not yet emergency braking. There is a chain reaction here, though. Each subsequent car will need to brake harder, until Car 5, which will have used up the 10-metre buffer before they even start braking. The cars will inevitably get closer together and, eventually, someone will hit emergency-braking territory and come to a complete stop, producing a phantom traffic jam.

The worst-case scenario is that a driver cannot brake enough and a collision occurs, which would definitely stop traffic. Such bad drivers are seldom sandwiched together at once, so, most of the time, collisions do not happen. However, this effect does show why only a few people braking too late, too hard, can cause a long phantom traffic jam.

Bottlenecks

Imagine you are driving down a two-lane stretch of road when you see a sign that one lane is closed ahead. Do you immediately indicate and politely change to the open lane and join the slowing queue to get through the narrow section? Or do you take advantage and storm down the emptier closing lane to the front of the queue, to prevail upon the kindness of a polite queuer to let you in? Is the polite thing to do necessarily the best, or do the queue jumpers have a point?

If everyone were polite, the queue would leave one lane empty up to the closure. This, in effect, means that the queue is twice the length it needs to be and, as driving in a queue is slow, increases the length of time spent queuing. It also enables the less polite drivers to skip to the front of a much longer queue, increasing the frustration of the polite ones.

What would happen if everyone remained in their current lane until the closure? This would make two queues, each half as long as in the first case, and each taking a similar amount of time to navigate. This means no one feels any road rage in taking turns merging at the obstruction. As the two lanes would also be travelling at similar speeds, merging is safer and faster, too. This 'zip-merging', as it is known, is the most efficient way of doing it.

So, the queue jumpers are, according to this reasoning,

selflessly trying to reduce the overall queuing time for their fellow commuters. In the UK – where queuing is a national sport and breaking queue etiquette is seen by many as a good reason to bring back capital punishment – it will take a lot to make us all adopt this new behaviour, even if it does save time.

Shortcut

Local knowledge is very useful, and many of us take advantage of shortcuts that other people aren't aware of in our area. Urban planners put a lot of effort into improving traffic flow, but sometimes opening up a new road to create a shortcut is counter-productive. Imagine a commute that, until recently, looked like this:

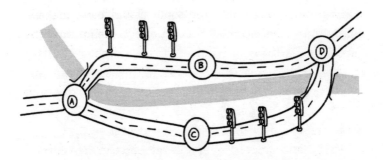

You would arrive at roundabout A and then choose whether to go north or south of the river. Thanks to your previous research into traffic flow, you know that the time taken to drive through the traffic-light sections (A to B and C to D) depends on how many cars are trying to get through, taking about a minute for every ten cars on the road. If c is the number of cars, then the time in minutes is given by:

$$time = \frac{c}{10}$$

The journey through the other sections of road (A to C and B to D) is less affected by the frequency of traffic and takes 25 minutes under normal circumstances, no matter how many cars there are.

During rush hour, you have counted that about 200 cars approach roundabout A, and as each route has similar roads, 100 travel on each side of the river. This means the traffic-light sections take 100 ÷ 10 = 10 minutes, giving an overall time from A to D of 10 + 25 = 35 minutes, whichever side of the river you choose.

Imagine that a new bridge is opened between B and C, allowing cars to quickly cross the river at this point. This gives the drivers going from A to D four choices of route: the two 'old' routes north and south of the river, ABD and ACD; and two new routes that involve crossing the river, ABCD and ACBD.

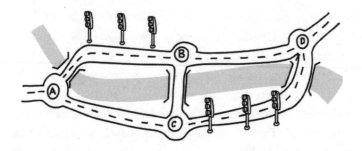

Using the times for each section as they stood before the bridge opened, you figure out the time for each route. We'll assume that it is very quick to cross the bridge and doesn't cost any time. ABD and ACD would be unchanged at 35

minutes. ACBD uses both of the slow sections and takes 25 + 25 = 50 minutes. No one in their right mind would choose that route. ABCD, however, goes through both of the quicker sections and takes 10 + 10 = 20 minutes. So drivers will start to take advantage of this new shortcut.

If you are the only driver to change to the new route, the A to B section still takes 10 minutes, as you are one of 100 drivers following their usual habit. When you cross the river and join the C to D section, there are now 101 cars going that way. 101 ÷ 10 = 10.1 mins. Your journey time is now 10 + 10.1 = 20.1 minutes. You have shaved virtually 15 minutes off your commute time and decide that you love the bridge.

A few days later and a few more people have caught on, thanks to the magic of social media and traffic reports. One morning, 100 drivers are using the ABCD route, while the remaining 100 stick to their ABD or ACD routes, 50 going north and 50 going south. This means that each traffic-light section has 150 cars travelling through, taking 150 ÷ 10 = 15 minutes. So, it takes you 15 + 15 = 30 minutes to go from A to D. This is still five minutes faster than before the bridge opened, but as more people get the idea the 20-minute run becomes a fond memory.

Meanwhile, the drivers on the old routes must now spend 15 minutes on the traffic-light sections, bringing their journey time to 15 + 25 = 40 minutes, 5 minutes longer than before. This gives them good reason to try out the new route.

By the end of the week, 150 of the drivers are using ABCD and 25 follow each of the old routes. This increases the time to get through the lights to 175 ÷ 10 = 17.5 minutes. The 'quick' route now takes 17.5 + 17.5 = 35 minutes, the

Anarchy Rules

The situation with the bridge shown in this chapter is an example of Braess's Paradox. This concept – that adding a new route to a network can slow the network overall – was first discovered by German mathematician Dietrich Braess (b. 1938) in 1968. As well as traffic, it can apply to a sports team where a star player that hogs the action acts like the extra road in the network and reduces the effectiveness of the team.

In real life, road planners in Seoul, South Korea, were surprised when traffic congestion improved with the demolition of a six-lane motorway that had originally been built to improve travel times in the city.

same time as before the bridge was built. The old routes now take 17.5 + 25 = 42.5 minutes, so these drivers have even more reason to try the new route.

Eventually, everyone is using the new route. All 200 cars going ABCD means that each traffic-light section takes 200 ÷ 10 = 20 minutes, giving a 40-minute total journey time that is 5 minutes worse than before the bridge opened. Mathematically, it would make more sense for everybody to pretend the bridge was shut and save themselves 5 minutes, but this would require a lot of trust. If fewer than 150 people use the bridge, those that do will save time and no one wants to sit in traffic longer than they must, especially during the rush hour.

You decide that your next research project may be bridge demolition.

CHAPTER 6

Science Fiction or Science Fact?

We have looked at the more mundane methods of getting around in this section, but even cars and bicycles were science fiction a hundred and fifty years ago. What can we look forward to in the future, and what are the mathematical rules that govern these space-age methods of transport? And are these going to be sustainable, environmentally friendly ways to travel?

Rocket Science

At the time of writing, if you want to travel long-distance in a reasonable amount of time, a jet plane is your only real option. But mankind first used rockets to get people into space in the 1960s. Reusable space shuttles were in use by the 1980s. Surely by now we should be travelling to Australia by rocket?

Planes are pretty fast – airliners can usually manage about 900 km/h. However, they are limited by the fact that they have to travel through air, which, although it provides the necessary lift on the wings and enables the engines to burn fuel, constantly drags at the plane. The air becomes too thin for most planes to be used above an altitude of about 20 km. The idea with rockets is that, after a powered launch phase gets you up to speed and into the thinnest air, the craft effectively flies the rest of the way without needing power, like a thrown ball. With little or no air in the way, the rocket can also travel much faster.

But doesn't it take ages to get all the way to space? No, is the simple answer. The distance from London to Paris is just a shade over 340 km. The distance from London (or pretty much anywhere else) to space is about 100 km. Yep, if you could drive straight up, you would find that space is only a couple of hours over your head. So, getting up to – and down from – space is only a small part of the journey.

Modelling Projections

Mathematicians, engineers and scientists have been studying projectiles – objects that have been launched or thrown – for a long time now. We know that projectiles travel in an arc known as a parabola, a family of curves that are easy to describe and manipulate mathematically. To model my rocket flight as a projectile, I'm going to make a few assumptions to make the maths a bit easier. Firstly, I'm going to assume that we can neglect the bit at the beginning and end of each flight in the thick part of the atmosphere where air resistance is a big deal. Secondly – and this will make the Flat Earthers happy

– I'm going to pretend that our rocket is flying over flat ground, unlike the Earth, which obviously (sorry, Flat Earth Society!) curves between London and Paris. Thirdly, rockets obviously take time to accelerate up to full speed, but I'm going to assume that it comes off the ground at full speed and then coasts for the rest of the trip, again like something that has been thrown.

The simplest parabola is given by y = x²:

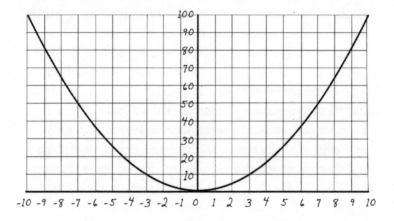

To plot this graph, I take the x values and multiply them by themselves to get the y values. This is cute, but it doesn't yet look like the path of a spaceship. To do that, I need to make some changes called transformations to the equation to get the shape I want. I need the graph to be the other way up and I need it to extend from the point (0,0) – London – and reach Paris 340 km away at (340, 0).

$$y = -\left(\frac{x}{17}\right)^2 + 20\left(\frac{x}{17}\right)$$

This gives me the London to Paris trajectory I want:

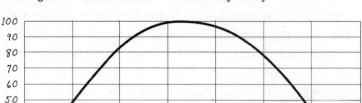

Prepare for Lift-Off

Now that we've sorted the flight plan, we can use it to work out our launch speed. One of the good things about modelling the rocket as a projectile is that we can treat how it moves vertically and how it moves horizontally as completely separate. The horizontal motion is not affected by anything (no air resistance, remember) whereas the vertical motion is only influenced by gravity. If we treat gravity (which won't get significantly weaker, even 100 km above the Earth) as constant then we can use the constant acceleration equations we met in the previous chapter for the vertical motion. The one we need is:

$$v^2 = u^2 + 2as$$

I'm going to be using this equation on the upward part of the flight. v is the final vertical velocity, which is zero at the moment the rocket reaches the top of the curve, when it changes from going up to going down. u is the initial vertical velocity at launch, which is what I want to know. a is the acceleration, in this case provided by gravity at -9.8 m/s² – negative because it pulls the rocket downwards. s is the vertical distance from the start point, 100 km (100,000 m), the edge of space.

Rearranging to make u the subject:

$$u^2 = v^2 - 2as$$
$$u = \sqrt{v^2 - 2as}$$

Substituting for the numbers:

$$u = \sqrt{0^2 - 2 \times -9.8 * 100,000}$$

This gives us a vertical launch speed of about 1,400 m/s. This is fast – the speed of sound is about 344 m/s, so this clocks in at about Mach 4, twice the top speed of Concorde, and we haven't even considered the horizontal speed yet. To find this, I need to know how long the rocket will take to get up to space. The v = u + at equation, which includes t representing time, will allow me to do this.

Rearranging to make t the subject:

$$t = \frac{v - u}{a}$$

Substituting for v, u and a:

$$t = \frac{0 - 1,400}{-9.8}$$

This gives t = 143 seconds, or 2 minutes 23 seconds. We can see from the graph that, in this 2 minutes 23 seconds, the rocket needs to cover 170 km (170,000 m) horizontally. We can use Speed = Distance ÷ Time as there is no acceleration involved horizontally:

$$\text{Horizontal speed} = 170{,}000 \div 143$$

This gives a horizontal velocity of just a shade under 1,189m/s. We can put the horizontal velocity and the vertical velocity together by using Pythagoras' Theorem, as they make a right-angled triangle:

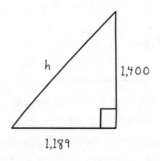

$$h = \sqrt{1{,}189^2 + 1{,}400^2}$$

Thus, we need the projectile to be launched at 1,838 m/s or about 6,600 km/h. The flight time would be 2 × 143 = 286 seconds, or 4 minutes and 46 seconds. London to Paris in under five minutes!

Our rocket liner, however, would not be a projectile. We can't launch things instantaneously at 6,600 km/h: we'd need time to accelerate up to that speed, as well as slowing down at the other end. Our projectile would need to be launched at an angle to 'hit' Paris, whereas a rocket would be launched straight up and manoeuvre into a trajectory. The Earth has a curved surface, too, which would need to be taken into account by the people working this out. It all gets a bit complicated – but hey, it's rocket science after all!

Long story short, if you want to get a rocket to Paris with intact humans inside it, five minutes is out. But our simple mathematical model shows that this approach could get us around the world much faster than is currently possible. It could also be done sustainably: it is possible to make rocket fuel from water that has been split into hydrogen and oxygen using green electricity. Once burned, this fuel becomes water again.

A Real Drag

If rocket travel were a thing, you'd still have to get to the spaceport. It would be ironic to spend hours getting there to spend minutes on the rocket to go halfway around the world. If Heathrow Airport near London became Britain's spaceport, I'd be doing well to get there, with the public transport currently available, in under four hours.

As we saw with rocket travel, air resistance is a real drag when it comes to getting around. Air resistance is proportional to the square of the speed you are travelling at, which means that if you double your speed, you quadruple the amount of air resistance. When you reach the top speed of a mode of transport it means that all the propulsive energy produced – whether it's your legs or a horse or an engine turning wheels or propellers – is going into beating air resistance.

Aerodynamic drag, the name for the air resistance force, is given by the equation:

$$F_D = \frac{1}{2}\rho v^2 C_D A$$

While this may look complicated, it's not too bad once you

get an understanding for all the symbols. C_D and A are to do with the geometry of the vehicle: A is for the cross-sectional area of the shape, while C_D is the coefficient of drag, a number that tries to account for the shape of the moving thing. Aerodynamic, streamlined vehicles have a lower drag coefficient. v is our old friend velocity, and it's squared which is the root of the problem I mentioned before. Rho, ρ, the Greek letter r, stands for the density of the fluid – in our case, air – that you are attempting to move through.

Deep Breath

The less air there is, the lower its density, so the lower the drag force and the more energy is left for going fast. A good example of this is the 1968 Olympics, which were held in Mexico City. At an altitude of 2,250 m, the air pressure is about 20 per cent lower than at sea level. Pressure is proportional to density, so the density of the air is about 20 per cent lower, too. This in turn means that the drag force is 20 per cent lower. As a result, world records were set in most of the athletics events under 800 metres, notably American Bob Beamon's 8.9 m long jump, which survives as the Olympic record to this day.

I don't want to have to travel to space every time I want to take advantage of this, so is there a way I can bring space down to Earth? Yes – if I have a vehicle travelling down a long tunnel, I could seal the tunnel and pump some or all of the air out, allowing the vehicle to travel faster.

I would not be able to use a combustion engine in this tunnel, as these rely on the oxygen in air to burn fuel. That's okay, though – electric vehicles are becoming more commonplace now and electric motors do not require air to

A Likely Story

Robert Goddard (1882–1945) was an American engineer credited with making huge practical leaps in rocketry. The Goddard Space Flight Center, where communications with the Hubble Space Telescope and the International Space Station are conducted, is named in his honour. Goddard, however, is also credited with inventing vacuum train travel in a short story. 'The High-Speed Bet' was never published, but *Scientific American* magazine featured the details of the story in an editorial called 'The Limit of Rapid Transit'.

produce power like internal combustion engines do. Now that the air is out of the way, less energy is spent pushing through it, so more energy can be put into spinning wheels.

Let's say I had a suitably air-tight electric car, capable of 35 m/s (120 km/h or 75 mph) in the normal atmosphere. How fast could it go in a tunnel with only 10 per cent of atmospheric pressure? Well, if I reduce the air resistance by a factor of 10, the top speed will theoretically increase by a factor of 10, making my car able to go 750 mph.

I say theoretically, because some of the energy of the car goes into the moving parts – the motor and wheels, driveshafts, etc. Traditional car engines have a range of revolutions per minute that they operate at, and they start to get unhappy at more than 6,000 rpm. Electric cars are often able to go up to 20,000 rpm. But 750 mph in a car is going to require much higher rpm, even with bigger gears. Is there a way to get rid of these moving parts?

Personal Magnetism

The answer to the last question is yes – using magnetic levitation and propulsion. You'll know from primary school that magnets have ends called poles: a north pole and a south pole. Opposite poles attract and like poles repel. Why? It's to do with the spin of elementary particles and the way electric fields produce magnetic fields, and vice versa. The word 'quantum' appears, which is scientific code for 'come back when you have a PhD'. Engineers with little understanding of quantum mechanics can, however, exploit the fact that like poles repel to make things float, including trains.

Not only that, but you can also use magnetism to push and pull a train down the tracks. Using rapidly switching electric currents, you can use the magnetic fields generated to pull the train towards a certain point and then push it as it goes past.

Maglev trains have been used since 1984, when you could take the maglev from Birmingham's train station to its airport. Now, most maglev train systems are in southeast Asia in China, Japan and Korea. Why aren't they everywhere? Well, despite being cheaper to run and more comfortable than ordinary trains, they are very expensive to build.

Maglev trains in vacuum tubes – called Hyperloop – don't exist yet, but watch this space. Several different companies are developing the technology and human trials have taken place. With any luck, you'll soon be able to travel to the other side of the planet in vacuum tubes and rockets, in only an hour or two.

PART 3

All in a Day's Work

You've arrived at work safe and sound,
thanks to the power of mathematics.
Now, how can you use it to smash your
working day and all that entails?
We look at hiring and firing, maximizing
profits and pairing people to their
ideal tasks.

CHAPTER 7

The Apprentice

Let's say you are a restaurant manager hiring new waiting staff. Your agency has sent a list of twenty candidates to interview, which is more than you have time for, and they all have similar experience and well-written CVs. If you interview someone, you must let them know whether they are successful or not at the end of the interview. Anyone you do not hire will find work elsewhere, so you do not get a chance for any call-backs. How can you give yourself the best chance of finding the best – or if not the best, a good – employee?

Let's take a look at some probabilities. The least time-consuming method would be to pick an applicant at random. The chance of hiring the best candidate here is one out of 20, or 5 per cent. Not great, but a chance some would be willing to take.

The other extreme would be to interview all twenty candidates, but because this would lead you no choice but to hire the last candidate, again the chance that this person is the best is just 5 per cent. You've effectively spent ages interviewing but aren't any better off.

The Strategy

In this situation, you have no idea how strong the field is before you start interviewing. You could rely on the applicants' CVs, but if theirs are anything like mine, they may not be 100 per cent legitimate. I mean, I did pro-actively and single-handedly install a new lighting system in the food preparation area within budget and ahead of schedule, but my wife would call this 'changing the kitchen lightbulb when it stopped working'.

One strategy would be to interview a throwaway sample of candidates who you will not hire, to get a feel for the calibre of the group. Then interview more candidates until you meet one who is better than everyone in the throwaway sample. This way you know you have a good candidate, if not necessarily the best.

You now need to consider how big to make the throwaway sample. In the diagrams below, each candidate's number shows their rank: 1 is the best candidate, down to 20, who is the worst. If the throwaway sample is too small there is a chance that you won't get any of the better candidates in the sample, and so the next best person will be mediocre.

Small throwaway
sample

You're hired!

Not interviewed

If you make the sample too big, you increase the risk of the best candidates being in the throwaway group, which means none of the remainder would be better and you'd be stuck with whoever tail-end Charlie happened to be. In fact, if the throwaway group contains the best candidate, you're doomed.

Large throwaway   Interviewed but   You're hired!
sample   rejected

Somewhere in the middle is an optimal number of candidates to have in the throwaway sample. How can we find that number? Let's reduce the number of applicants to some low numbers to see how this would play out.

With one applicant, that person is simultaneously the best and the worst candidate and no maths is required to see this. With two applicants, random guessing gives a 50 per cent chance of getting the best person. Our throwaway sample could only have one applicant in it, who is either the best person or the other person, so again we have a 50 per cent chance of getting the best.

So far, the strategy does not improve on random selection, but we'll see that it comes into its own as the number of applicants increases. With three applicants, your random guess is now at 33.3 per cent. Using the strategy, you could have one or two throwaway interviews. By now, we can see that the order you happen to see the candidates in is really important. If each applicant has a rank of either first, second or third, you can look at the various permutations of the orders you see them in to work out your chances of success. Three applicants could be arranged in six ways:

1 2 3; 1 3 2; 2 1 3; 2 3 1; 3 1 2; 3 2 1

Let's see how many times you'd get Applicant 1 if you had one throwaway interview. For 1 2 3, you throw away the best person, and end up taking 3 as 2 would fail to impress compared to 1:

Throwaway sample

Interviewed but rejected

You're hired!

Not so great. For 1 3 2 you would at least get the second-best applicant:

Throwaway sample

Interviewed but rejected

You're hired!

I hope you're getting the idea of how it works now. For 2 1 3 and 2 3 1 you would throw away 2 and end up hiring 1 in both cases – nice! 3 1 2 nets 1 again, while 3 2 1 gets 2.

Overall, this method gives 1 three times out of six – a 50 per cent hit rate, quite a bit larger than the 33.3 per cent of random chance. If you interview two people in the throwaway group, you can see that you are back to hiring the last applicant whoever they are, so you are back to a 33.3 per cent chance if you do this.

Let's move up to four applicants. There are twenty-four ways of ordering the candidates now. I'll list the outcomes for four applicants with one in the throwaway by underlining the winner:

1234; 1243; 1324; 1342; 1423; 1432; 2134; 2143; 2314; 2341; 2413; 2431; 3124; 3142; 3214; 3241; 3412; 3421; 4123; 4132; 4213; 4231; 4312; 4321

This gives us 1 eleven times out of 24, giving a 45.8 per cent hit rate. I'll spare you the details, but if you have two throwaway interviews you end up with 1 ten out of 24 times or 41.7 per cent. Again, for three throwaways you have to take the last candidate, which is the same as just guessing in the first place: 25 per cent.

We can see that having no candidates in the throwaway group and having all-bar-one in the throwaway group gives us the same chance of success as random chance. Somewhere in the middle is the best size for the throwaway group, giving the best chance of getting the best staff for your business.

Twenty's Plenty

For our scenario of twenty candidates, there is a lot of calculating to do. If you copied our system from above, you'd need to write out all the possible orders of twenty applicants. There were six orders for three applicants and twenty-four orders for four. For twenty applicants there are 2,432,902,008,176,640,000 possible orders, a bit more than 2.4 quintillion. Then you'd have to work out the successful candidate for each size throwaway group from zero up to nineteen.

Fortunately, you can apply a bit of maths know-how and a spreadsheet to speed things up a bit. If you imagine your candidates in a queue, you can work out the probability that a candidate in a particular position will be successful.

We want to work out the probability of the interviewee with the question mark being selected. If the throwaway sample includes their position (shown by the dashed box

---

Suitorbility

This problem, originally called the Fiancée Problem by American mathematician Merrill Flood (1908–1991), has also been called the Sultan's Dowry Problem, the Fussy Suitor Problem and the Googol Game.

It's probably worth noting that if the applicants themselves discover how your selection process works, they may be unwilling to come for the interview! You would also, I suspect, be in breach of employment law in most countries.

on the diagram below), there is no way that candidate can be selected, so their probability of success is zero.

If they are outside the throwaway sample, for them to win they need someone in the sample to be better than everyone in the queue before them.

In the example above, we need one of the people in the throwaway group to be ranked higher than the two preceding our guy, without out-ranking him.

The chance of this happening is given by the size of the throwaway sample divided by the number of people in the queue before the candidate we are considering. In the diagram above there are six people in the throwaway group and eight before our guy, giving a 6 ÷ 8 = 75 per cent chance of our guy being in the winning position. Finally, the chance of our guy actually being the best candidate as well as the winning candidate would be one in twenty which is 5 per cent. If you multiply the probability of them winning by the probability of them being the best, you get the probability of that scenario bearing fruit. Below is a graph of the probability of getting the best candidate for each throwaway sample size, which I've calculated using the method above:

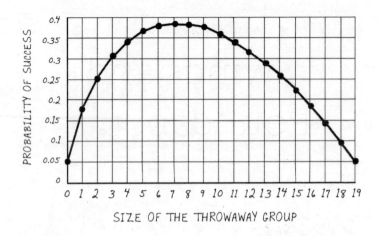

You can see that the curve peaks at a sample size of seven applicants, which would give us a 38.4 per cent chance of success, well above the 5 per cent chance of getting lucky by guessing. In general, the maths says that in this type of situation, whether you are choosing a waiter, employee or your future spouse from the dating pool, you should meet 37 per cent of the people, disregard them, and then choose the next one that you encounter that is better than all the ones met so far.

CHAPTER 8

Swings and Roundabouts

Many businesses make a variety of products which they then sell to make a profit. The concept is simple, but the devil is in the detail. How many of each product should you make to maximize your profits? This is where mathematics can help.

Embrace Inequality

The first problem to overcome in this situation is to translate the business into a mathematical model. Let's use an example. Imagine a company, Swings & Roundabouts Limited, which fabricates – you guessed it – swings and roundabouts. There are certain constraints on their business:

1. They get enough raw materials delivered to build up to 10 swing sets and 10 roundabouts each day.
2. Contracts for the workshop team state that they must build at least 7 products per day.
3. Union rules stipulate that the workshop team need only fabricate up to 15 products per day.
4. Swings & Roundabouts Limited makes a profit of £40 on swing sets and £100 on roundabouts.

This is a very simple problem which you may be able to solve in your head, but once you understand the concepts you will be able to see how it could be extended to much more complex situations. To start with, let's introduce two variables: the number of roundabouts built each day, r, and the number of swing sets, s.

The first bullet point tells me something about r and s – that they must both be less than or equal to 10:

$$r \leq 10$$
$$s \leq 10$$

The second bullet point tells me that the total of r and s must be greater than or equal to seven:

$$r + s \geq 7$$

The third bullet point means that the total of r and s must be less than or equal to 15:

$$r + s \leq 15$$

To deal with the last bullet point, I need another variable to stand for the profit made: P.

$$P = 100r + 40s$$

When in Doubt, Draw it Out

If you have a sinking feeling that you're going to have to somehow solve all the inequalities above, don't worry. Our method is going to involve a visual solution. This means we have to represent all the inequalities above on a graph. It's easier to start with equations r = 10 and s = 10.

If the horizontal axis represents r, then a vertical line going up from 10 shows all the places where r = 10. Likewise, if the vertical axis represents s, a horizontal line extending from 10 shows all the places where s = 10:

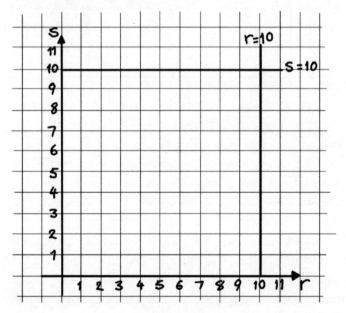

The inequalities, rather than being represented by a line on the graph, are represented by a region. To show this, I will shade in the areas that are not allowed: that is, anywhere outside the square shown above, because this is where r and s are larger than 10:

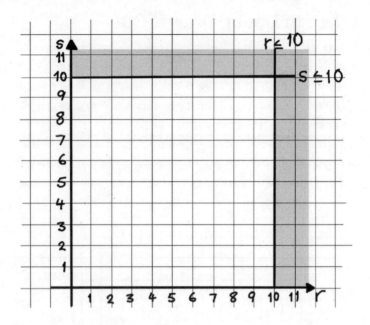

The unshaded area is called the feasible region, because any point on or inside the square obeys the inequalities. But we have more inequalities to add before we get too pleased with ourselves. As before, I'll look at equations and then shade regions to make the inequalities. For r + s = 7, I need to think of the points where r and s have a sum of seven: (0,7); (1,6); (2,5);(3,4); (4,3); (5,2); (6,1); (7,0). These make another straight line on the graph. Again, I shade the area where r + s make less than seven as these are not allowed. I can follow a very similar process for r + s ≤ 15 to work out all the places where r + s = 15, draw a line and then shade:

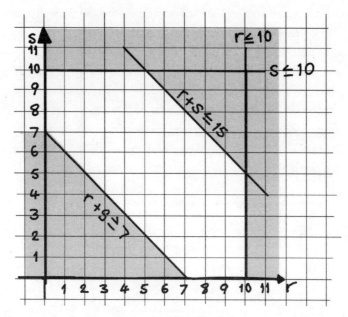

The unshaded region shows all the points that are permissible by the inequalities. If I pick any point in or on the sides of the hexagon, it will follow all the rules from the first three bullet points. What remains to do is to work out where the maximum profit lies.

Profit Prophecy

I could take every point in the hexagon, work out the profit, and see which was largest. There are seventy-eight points, so that's a lot of calculations. Instead, I'm going to pick a point within the feasible region, let's say (5,6). This is where r = 5 and s = 6, so if I feed these into our profit equation I get:

$$P = 100 \times 5 + 40 \times 6$$
$$P = 500 + 240$$
$$P = 740$$

So that combination of swings and roundabouts will make the company £740 in profit. There will be other places that will make £740 profit, though. For instance, it's fairly obvious that if I make r = 7 and s = 1 this will give the same amount. I can build up another line on the graph that will show everywhere P = 740. Many of the points on this line aren't possible in reality, because only whole numbers of swings and roundabouts can be produced. I can also work out the line for making the profit £900:

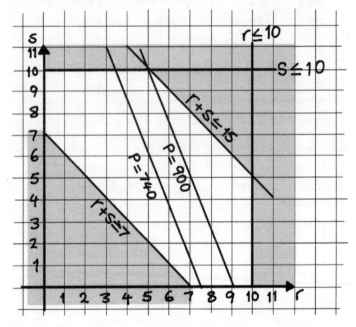

I notice something here: the P = 740 and the P = 900 lines are parallel. As I increase the profit, the value of P, the line moves across the graph. If I keep increasing P, eventually the line will be entirely outside the feasible region. But I can see that the last feasible place that would be included

Cooking the Books

Benford's law, or the First-Digit law as it is also known, states that the first digits of collections of numerical data (such as accounting figures or vote counts) are not evenly spread. As there are nine digits that could be first, you might expect each one to occur one-ninth of the time. However, the lower the number, the more likely it is to be a first digit: one is a first digit over 30 per cent of the time for data that obeys this law.

Benford's law, named after American physicist Frank Benford (1883–1948), is used to analyse figures reported for tax purposes – any anomalies may be a result of fraudulent accounting. So if you are going to cook the books, make sure you do it mathematically!

is (10,5). This means r = 10 and s = 5, ten roundabouts and five swing sets, give the maximum profit of 100 × 10 + 40 × 5 = £1,200. Problem solved!

CHAPTER 9

Perfect Partners

Now that you've worked out how to maximize your company's profit, you need to get the products made. Imagine that we had five different workstations and five employees. Because you believe in ongoing professional development, the employees can work at more than one station. Is there a straightforward way to get everyone to a station and start producing swings and roundabouts?

Match Made in Heaven

Let's get some details. Your five workers are Anna, Bill, Carla, Danny and Elspeth. Anna can work at stations 1 and 4; Bill stations 2, 4 and 5; Carla stations 3 and 4; Danny stations 1 and 3; and Elspeth stations 1, 3 and 5. I don't know about you but my brain is already starting to hurt.

On a particular morning, Bill is at station 4, Elspeth's on 3 and Anna is on 1. I can show this using something called a bipartite graph, which shows the people on the left and the stations on the right. I've drawn lines to show where people and stations are already matched.

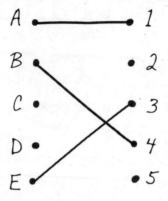

I now follow a simple recipe – or algorithm, as mathematicians like to call them – to match everyone up. I choose one of the unassigned people – in this case, Carla or Danny – and get them to choose a station. If there's no one there, great, move on to the next unassigned person. If there is someone there, they choose another station they can operate. Again, if there's no one there, go to the next unassigned person, if there is they choose another station. And so this cycle repeats, until everyone is assigned. There may be more than one way to match everyone up, so the choices the individuals make can lead to different matchings. This algorithm is called the Maximum Matching algorithm.

Let's say I start with Carla. She is trained on stations 3 and 4, both of which are filled. She chooses 3, which bumps Elspeth. Elspeth decides to go to station 5, which is empty. The graph now looks like this:

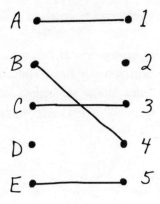

I can write the sequence of changes like this, using = to mean 'is matched with' and ≠ to mean 'is no longer matched with':

$$C = 3 \neq E = 5$$

This is called an alternating path, a convenient way of showing your working. That leaves Danny. Danny can work at 1 or 3, both of which are filled. He chooses 1, bumping Anna, who chooses 4. This bumps Bill, who chooses 2 which is empty. This alternating path would be written as:

$$D = 1 \neq A = 4 \neq B = 2$$

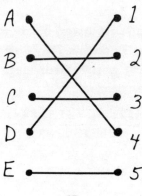

Everyone is now assigned – which is called a Maximum Matching.

If this situation seems trivial, imagine that you had a large workshop with hundreds of stations and workers. Then, having a convenient algorithm would be a huge help, and software could do the matching for you.

Bin It

Meanwhile, the delivery team are trying to sort out consignments of swing sets into vans to deliver to local garden centres and DIY stores. The company has three vans, each of which can carry sixty swing sets. Each customer has requested a different number of sets. The question is: can we pack each van so that no order has to be split between two of them?

Situations like this are known as Bin Packing problems and there is no quick way to find the perfect solution as yet. But there are a few mathematical ideas we can employ to help us find a solution, even if it is not necessarily the best one.

Here are the shipments, listed in order of purchase:

9, 14, 25, 13, 26, 8, 23, 28, 12, 22

The first thing I can do is to work out the minimum number of vans I will need. To do this, I work out how many swing sets need delivering in total and divide by 60, which is how much each van can carry:

$$9 + 14 + 25 + 13 + 26 + 8 + 23 + 28 + 12 + 22 = 180$$
$$180 \div 60 = 3$$

Taken to Extremes

While this problem becomes more difficult as the numbers get bigger, if you head up all the way to infinity, they melt away. Even if you had an infinite number of vans that were already full, you can actually pack infinitely more shipments into them. How? Simple. Shift all the contents of van 1 into van 2, from van 2 to van 4 and from van 3 to van 6 and so on. This process will leave all the odd numbered vans empty, and there are infinitely many of these: plenty of room for your shipments.

If this sounds weird, you can blame German mathematician David Hilbert (1862–1943), who demonstrated the strange properties of infinitely large sets, as well as setting the twenty-three 'Hilbert problems' that are still keeping mathematicians busy to this day.

This tells me that I need exactly three vans, if each one is completely full. One way to pack the vans is called the First Fit algorithm: I simply load the vans in the order above, putting the shipment in the first van with space for it. This is a simple approach but is unlikely to work perfectly.

I can fit the first three orders into van 1, because 9 + 14 + 25 = 48, leaving space for 12 more sets. I can't fit the next shipment of 13 in this van as it will take me over 60 sets, so I put it in van 2. 26 will also have to go in van 2, but 8 will fit in van 1, bringing it up to 56. 23 won't fit in van 1 or 2, so I put it in van 3. Same story with 28, bringing van 3 up to 51.

12 can fit in van 2, bringing it up to 51 too. Unfortunately, the 22 can't fit in any van, so will have to wait for another delivery run. This algorithm has allowed me to deliver 56 + 51 + 51 = 158 out of 180 swing sets.

A better algorithm is to order the shipments into decreasing order of size and apply the First Fit to that list. This is probably how you would pack the boot of your car with suitcases – you put the biggest ones in first and then fit the smaller ones around them. I'll spare you the blow-by-blow and show you the results in a table:

Van	Consignments	Total
1	28 26	54
2	25 23 12	60
3	22 14 13 9	58

This time, I had to leave out eight swing sets, but I did manage to fill one van completely. It's still not perfect, but the only way to find a perfect solution is to either use a computer to work out every possible combination, or to try to do it by eye. This isn't too hard when you only have ten deliveries to consider, but if you were a major parcel delivery company with millions of packages it would take a lot more effort.

PART 4

Retail
Therapy

Whether it's business or pleasure, buying
things requires a lot of numbers know-how.
In this section we look at everything from
the mathematics of money, to how to win at
eBay and how to deliver the goods.

CHAPTER 10

In for a Penny

Wealth has long been a preoccupation for the human race. One of the first ways of making your wealth more portable was to carry around chunks of precious metals. Over time, these developed into systems of coinage, usually backed by a state. We no longer carry around coins constituted from materials worth their face value. Instead, that value theoretically represents a pile of precious metal in a bank vault somewhere. The same with bank notes: British ones are emblazoned with 'I promise to pay the bearer on demand the sum of' whatever the value of the note is.

Freshly Minted

Ever since there have been coins and banknotes there has been a war between the counterfeiters and the people issuing the money. In 1696 Britain devoted one of its finest minds – that of Isaac Newton – to sorting out its currency.

When Newton first took charge of the Royal Mint, it was estimated that as many as 10 per cent of British coins were forgeries. Newton was charged with tackling this problem, as well as improving the consistency with which coins were produced in the first place. He turned his renowned scientific zeal to resolving these issues and was in charge of the mint until his death in 1727.

Although coins have come in all shapes and sizes, these days it is useful if coins roll well. This allows them to be used in coin-operated machines, such as vending and ticket machines. Circular coins are a relatively easy shape to counterfeit, however, so many countries have done some geometry to make non-circular coins that still roll nicely.

Roll with it

If you imagine a coin rolling along a flat surface, there is always the same distance from where the coin touches the surface to the top of the coin:

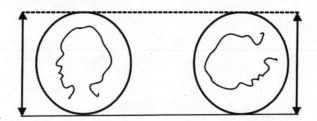

If you imagine a triangular coin (such as the $2 one issued by the South Pacific Cook Islands in 1987). It would be difficult to get this to roll for long as the distance from the surface to the top of the coin varies a lot:

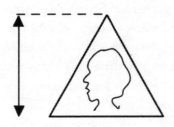 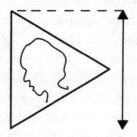

Circular things roll well because they have constant width – that distance from the surface to the top of the coin. Triangular ones do not roll well because they do not have constant width and the width varies a lot. But is it possible to make shapes other than circles with constant width?

If I make the triangle curvier, it could work. I draw a circular arc from each corner to the opposite side of the triangle, with a radius the same as the side of the circle:

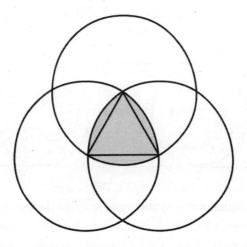

The central shape is known as a Reuleaux triangle:

This shape, invented by German engineer Franz Reuleaux (1829–1905), always has the same distance from top to bottom when on a flat surface, like a circle. So a coin this shape will roll nicely in a coin-operated machine:

It has other advantages, too. It would be easy to distinguish by touch by people with visual impairments. It is also very distinctive: Bermuda issues coins this shape to commemorate special occasions.

The Reuleaux triangle has a couple of other interesting properties. As it has constant width it is a good shape for manhole covers, as it is not possible to drop the cover through the hole accidentally. Also, it uses less material that a circular cover of the same width, making it cheaper to produce.

You may be thinking 'why don't we make wheels this shape, then, if they would do the same job as circles, but with less area?' The reason is that, unlike the circle, the centre of the triangle does not stay in the same position when it rolls, so a Reuleaux wheel would give a very bumpy ride. That said, Guan Baihua from China invented a bike with special axles using Reuleaux shapes for wheels in 2009.

If you make a drill bit in an off-centred Reuleaux triangle shape you can get it to drill almost-square holes, a fact exploited by Panasonic to build a robotic vacuum cleaner that can reach into corners.

More Than Three Sides to the Story

You can construct other shapes with more sides using the same approach. British readers will, of course, be familiar with the Reuleaux pentagon and heptagon: our 20 pence and 50 pence coins are these shapes. You can make a Reuleaux polygon from any shape with an odd number of sides.

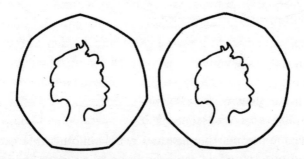

The original British pound coin, first issued in 1983, was circular and it was estimated that about 3 per cent of pound coins in circulation were counterfeit. In 2017, a new dodecagonal (12-sided) coin was released, made from two metals to prevent forgeries. This coin is not a Reuleaux shape because it has an even number of sides and anyway, with the larger number of sides, adding the circular arcs makes very little difference to the coin's ability to roll.

Another Dimension

If we head into three-dimensional territory, it is possible to make solid shapes of constant width. The sphere is the obvious one, but it is also possible to use the logic behind the Reuleaux triangle to make a shape called a Meissner tetrahedron – a curvy triangular-based pyramid shape that always has the same height from base to top.

Take Note

We've seen some geometric strategies for coins, but what about banknotes? That's where the real money is.

Notes, being more valuable, have a variety of counter-counterfeit measures in place. British notes are now made from a plastic polymer rather than paper and employ fine printing, see-through windows, colour-changing borders, silver and gold foil, raised print and hidden ultraviolet markings. They also use holograms, those seemingly magical pictures that change before your eyes, depending on the angle you look at them.

Holograms are essentially a photograph but taken with laser light rather than normal visible light. As we saw in Chapter 1, laser light is all of one wavelength and all those waves are in step. To give an analogy, normal light is like throwing a bunch of stones into a pond – the waves spread out in all directions, are different heights and lengths, and interfere with each other when they meet. Laser light is like equally spaced waves in the sea, all heading in the same direction.

Surreal Image

In 1973 the Surrealist artist Salvador Dalí and the shock-rock performer Alice Cooper collaborated to produce one of the first holograms used as art. The work, titled 'First Cylindric Chromo-Hologram Portrait of Alice Cooper's Brain', features Cooper wearing several million dollars' worth of jewellery, singing into a statuette of the Venus de Milo, in front of a sculpture of his brain topped with an ant-covered éclair.

Obviously.

Laser Vision

To make a hologram, the first step is to split the laser light in half using a special mirror. Both beams are then passed through lenses to spread them out. One beam, called the reference beam, is projected directly onto the holographic film. The other, called the object beam, is bounced off the object and then onto the holographic film. Just like taking a normal photo, the whole set up needs to be free of vibration, but the holographic process needs to be very still, as a passing car or even someone walking nearby could be enough to blur the exposure.

So, the reference beam and the object beam are added together to create what the holographic film records. I could write this as:

$$R + O = H$$

Whereas a developed photographic film looks like the image recorded, the holographic film does not. Instead, the holographic film records what is called an interference pattern. To see the object again, I need to see the object beam. I can do this by rearranging the formula above:

$$O = H - R$$

The '-R' can be achieved by shining the same laser light I used to make the hologram through the holographic film, but from behind. This will display the image to anyone in front of it in what is called a transmission hologram. These have been used in museums displays and art installations. Banknotes use reflection holograms, where the light comes from the same side of the hologram as the viewer but bounces off the holograms to produce the image.

This process means that when you view the finished hologram, you see what the laser 'saw' from the same angle. So as you change the angle that you view the hologram from, you are only able to see the object as if viewed from that angle. This gives the amazing sense of three-dimensionality that makes holograms seem so magical.

From the description above, you can see that making holograms is not easy at all. This is what makes them suitable for authenticating objects such as banknotes, passports and other identity cards, as well as other products that are often faked, such as DVDs.

CHAPTER 11

Next-Day Delivery

Shopping has never been easier. From the comfort of my sofa, smartphone in hand, I can browse and buy pretty much any product imaginable, from food to musical instruments, to holidays, to stocks and shares. As online retailers don't need to spend money on physical stores, their products are often temptingly cheap, too. Unlike physical shopping, though, these products need to get from their location to mine, and often the delivery times and charges are big factors in our purchasing choices.

The problem of delivery also affects physical stores, which need deliveries of products to keep the shelves stocked. Hence the science known as logistics – getting stuff where it needs to be, when it needs to be there.

Child's Play

Moving stuff around a network of roads, railways, airports and dockyards is an extremely complex job.

Mathematicians have been working on it for centuries: the field of graph theory tries to solve these problems. You may not have heard of it, but you've almost certainly encountered it, albeit unknowingly.

When I was a child there were quite often sections in puzzle books with the challenge of copying a shape without lifting your pencil off the paper or having to draw a line twice. This shape was popular:

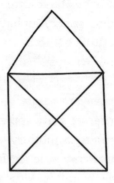

With a little experimentation, you'd discover that you could draw the shape if you started on the either of the bottom corners. If you start anywhere else, it won't work. Can we work out why? Well, if I mark all the points where the lines meet and count how many lines meet there, I get this:

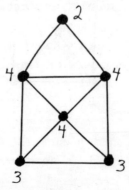

I notice that all the points – called nodes or vertices in graph theory – have an even number of lines meeting at them, apart from the bottom two. Now, think about your pencil travelling around this diagram – or graph, as mathematicians call it. If your pencil visits a node, it has to go in and then out, so you need two lines going to the node. If it goes through it again it needs another two lines, so we need an even number of lines in order to visit it several times. The bottom nodes have three lines, which implies we can travel through it once, but have to either start there or stop there. This turns out to be the case: if you start on one of the odd nodes, you finish at the other. If I don't start at one of the odd ones it becomes impossible, because I can't meet this criterion any more.

There you go – graph theory applied to a child's puzzle. We can extend the logic to another couple of cases. If I had another graph where every node was even, I should be able to start anywhere and draw the graph without lifting my pencil. What is more, I will always finish up where I started. Try this one:

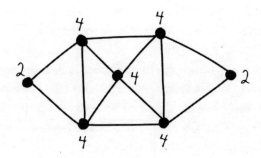

The opposite is true, too. If I have a graph with more than two odd nodes, it is impossible to draw without lifting your pencil or repeating a line:

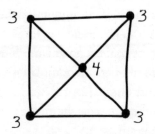

Pathway to Success

What has this got to do with my online purchase? Well, it's not a huge leap to imagine that the graphs we just looked at could represent a travel network, with the lines representing roads and the nodes representing delivery points.

But our discoveries so far are more useful for the post office than an online retailer. In the child's puzzle we travel down each road once but visit each node several times. A completely even graph would suit a postal worker perfectly as they can travel down each road delivering letters and return to their starting point – this is even called the Chinese Postman problem. The delivery driver, however, wants to visit each node just once and go the minimum possible distance around the graph, known as the Travelling Salesman problem. Can we use our new knowledge to help the driver?

Not easily, is the answer, and we'll see why in a bit. First, we need to think about cycle lanes.

Imagine that you are a road planner and you've been

tasked with connecting a number of towns by adding cycle lanes along the sides of existing roads. It needs to be possible to get to any town purely on the cycle lanes, but you don't need to connect every town directly. It would be great for the budget if you could use the shortest possible distance overall.

I can draw a graph of the towns as we have before. This is a bit like the map of the London Underground, showing connections rather than geographic positions. As we are also interested in distances, we need to see these on the graph, shown in kilometres:

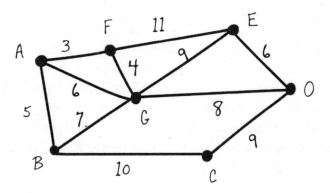

American mathematician and computer scientist Joseph Kruskal (1928– 2010) came up with a very simple way to solve this problem. He spotted that you can build up the route by using the shortest roads until you have joined every node. The one thing you must avoid is making any triangles, because if you have three places you only need two roads to connect them: the third one is superfluous.

So, looking at the map above, I can see that AF is the shortest road at 3 km, then FG and AB at 4 and 5 km respectively. There are two 6 km roads, AG and ED, but AG would form a triangle with AF and FG, so we'll go for ED instead. Next is BG at 7 km, but again this would form a triangle, so I ignore it. With the addition of GD and DC, I have connected the towns with the shortest possible route:

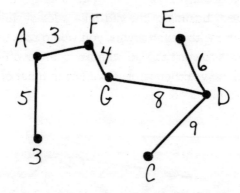

Mathematicians call this a Minimum Spanning Tree. I suppose it does look a little like a tree if you're a mathematician who doesn't get out much. You can tell the council you'll need 3 + 4 + 5 + 6 + 8 + 9 = 35 km of bike lane to do the job.

A Long Way to the Top

This doesn't solve the problem for the delivery driver, but it gives us some idea of how far they must travel. If they were delivering to the same network of towns, we now know that it would be possible to deliver to all of them by travelling each of those roads twice: a total distance of 70 km. This isn't the shortest route, but it would be viable. As we continue our work, we'll be trying to drop this figure of 70.

Sadly, there is no algorithm or quick method to work out the shortest Travelling Salesman tour of a graph. The only way to guarantee finding the quickest route is to look at every single possible route and see which is shortest. For our network of seven towns, we could probably do this quite quickly, but if you were a large national or international business, it could take a lot of time. In the general case with seven towns, you would pick one of the seven towns to start at, then choose from six others to go to next, then five others and so on. The number of possible routes is then:

$$7 \times 6 \times 5 \times 4 \times 3 \times 2 \times 1 = 5,040$$

Mathematicians have a shorthand for multiplying like this: 7!, called seven factorial. While this doesn't seem like too many for a computer to handle, the typical delivery driver makes about 200 stops per day:

200! = 200 × 199 × 198 × .... × 2 × 1 = 78865786736479050355 23632139321850622951359776871732632947425332443594 49963403342920304284011984623904177212138919638 83025764279024263710506192662495282993111134628572 70763317237396988943922445621451664240254033291864 13122742829485327752424240757390324032125740557956 86660226031904170324062351700858796178922222789623 70389737472000000000000000000000000000000000000 000000000000000000

There is no name for this number. It is 375 digits long and has 49 zeroes at the end.

Yet, somehow, we get our parcels delivered: over 7 million every day on average in the UK. How do they do it?

Eureka Moment

While we don't have a quick way of finding the best solution to the Travelling Salesman problem, we do have ways to find other decent solutions. Mathematicians refer to these as heuristic algorithms. It's exactly what you do if you choose a route using a road atlas – you pick a route that will get you there, that could probably be improved if you had local knowledge of shortcuts and traffic, but is still acceptable. It does the job.

Eureka

You may have heard the story of Archimedes' (287–212 BC) sudden moment of clarity in the bath. He realized the relationship between the volume of an object and the water it displaces, which allowed him to solve a tricky problem set to him by his boss, King Hiero II of Syracuse, no less. In his excitement, he leaped out of the bath and ran down the street naked, shouting 'Eureka!', Ancient Greek for 'I have found it!'

Well, our word 'heuristic' has the same Greek root and idea behind it. A heuristic method is one which gives a viable solution to a problem in a reasonable amount of time. As we saw above, working out all the routes in a Travelling Salesman problem to find the best one would take much longer than the time you would save in just eyeballing the map and picking a reasonable route instead. So next time you are trying to explain a decision you have made, say you used a heuristic method and see if anyone objects.

One of these heuristic algorithms is called the Nearest Neighbour algorithm. In case you can't guess from the title, it works by picking a starting place and then going to the nearest unvisited node on the graph. When you've visited every node, you find your way back to the start by the shortest route.

If a delivery driver wanted to start at town A, the nearest neighbour is F. F's next nearest neighbour is G. Following the algorithm would take me then to B, C, D, and E. From E we look for the shortest way back to A, which means going via F:

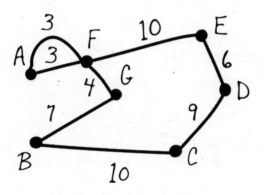

This gives the driver a route of 3 + 4 + 7 + 10 + 9 + 6 + 10 + 3 = 52 km, a decent improvement on the 70 km we worked out before.

Human Heuristics

Human beings are good at solving things heuristically. If I wanted to deliver to all the towns on the map we looked at before, without working out thousands of routes and comparing them, or using the nearest neighbour, I could start with the Minimum Spanning Tree and see whether I can tweak it into more of a circular route. I could swap the GD road for GE, adding one extra kilometre to the tree. Then going from C to B would give me a circular delivery route:

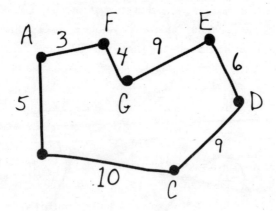

I know the distance here is 35 - 8 + 9 + 10 = 46 km, six kilometres less than the nearest neighbour got me. It looks like humans aren't completely out of the loop. Yet.

CHAPTER 12

Going Once

Many of us enjoy shopping, whether online or in person. Retail therapy can be a great way to unwind and relax. But sometimes we want a thrill, and this is where auctions come in. At an auction, you bid, and it takes effort, skill and luck to win whatever it is you are bidding for. There is exhilaration in winning and disappointment at losing. It turns buying stuff into a competition.

This excitement brings millions of people to online auction sites like eBay, keen to grab a bargain and experience the thrill as the clock ticks down while you are the highest bidder. But can mathematics help you do this more effectively?

Horses for Courses

EBay has set things up perfectly. You are in the market for a product. You search for it and are presented with a list of choices. You scout through for the right one, decide what value it has for you, type in a bid and off you go. EBay automatically puts in the lowest bid that will put you in the lead. You can go away happy that, if the price is less than your maximum bid, you will get a bargain. If it is above your maximum bid, it's more than you wanted to pay so you can bid on another similar item. Easy. Everyone should just put in the maximum they are prepared to pay and we can all get on with our lives.

But it seldom goes like that. There are a few factors at play here, but the main ones are: many people don't actually understand how eBay's bidding system works; eBay wants every auction to produce as high a price as possible, because that's how they make money; and human psychology.

There are, broadly, three kinds of bidding methods: max(-imum) bidding, incremental bidding and sniping. Max bidding is what I have already described – put your maximum bid in early and see what happens. Incremental bidding is where you enter prices close to the current winning bid and keep going until you are either in the lead or reach as much as you are willing to pay. Sniping refers to entering your maximum bid in the last moments of the auction.

The Theory

Life is competitive and no one likes to feel they are missing out. The study of these phenomena is known as game theory, which is the branch of mathematics that deals with the potential outcomes of these sort of situations and the value of winning to the individuals involved. You are using game theory when you play games such as chess or rock paper scissors, but also when you decide which queue to join at the supermarket, choose which shares to invest in or ask your boss for a salary raise.

The original problem posed in game theory is known as the Prisoner's Dilemma. Two criminals have been caught and are being interrogated separately by the police. The police have evidence to convict each of them of a minor crime, with a prison sentence of one year, but not enough to convict them of a more serious crime with a three-year sentence. If a criminal provides evidence against their partner, they will give the police what they need to convict the partner of the more serious crime and will reduce the sentence of the helpful criminal by a year. The criminals are isolated and cannot communicate. What should they do? What would you do?

Mathematicians show the information in a table called a pay-off matrix:

	Prisoner 2 loyal	Prisoner 2 disloyal
Prisoner 1 loyal	Both serve 1 year	Prisoner 1: free Prisoner 2: 3 years
Prisoner 1 disloyal	Prisoner 1: 3 years Prisoner 2: free	Both serve 2 years

This is where the human psychology comes in. The best course of action for the two criminals collectively is to both stay quiet and serve one year each. If one is disloyal and the other not, three total years are served. If they are both disloyal they get two years each, giving a total of four years:

	Prisoner 2 loyal	Prisoner 2 disloyal
Prisoner 1 loyal	2 years	3 years
Prisoner 1 disloyal	3 years	4 years

These are criminals we are talking about. Even if they had made an honour-among-thieves-style pact beforehand, can they trust each other? Can they risk it? The chance of personal gain (not going to prison at all) makes it most likely that the criminals would both be disloyal and end up with the worst outcome for the pair, four years in total – the price of anarchy, as we saw in Chapter 5. The criminals – and most other people, too – would play it safe and look at the worst-case scenario for each case. If a criminal remains loyal, the worst-case scenario is three years in prison. If they are disloyal, it is two years in prison. So, the best worst-case comes from being disloyal. It's not called a dilemma for nothing.

The Prisoner's Dilemma applies to many other situations: countries paying to reduce their CO_2 emissions, professional cyclists taking performance-enhancing drugs, or even when to call your parents without a guilt-trip.

Value for Money

In the Prisoner's Dilemma, the value of 'winning' or 'losing' is pretty obvious. When you are bidding for an item on eBay, the value is less obvious, and changes from person to person. Imagine an auction for a teacup with a nominal value of £5. For someone who just wants something nice to drink their tea out of, they might pay up to but not beyond this value. If they miss out on the auction they'll just bid on something else, or maybe just order something with no auction.

Someone else has been collecting this set of china for the last ten years. This is the last piece they need and, although the china is not particularly valuable, it is quite rare. They might be willing to pay significantly more than £5 for it. This would make the seller, and eBay, very happy.

There is also the psychology of the starting price. You may be selling an item that will go for over £100. If you start the bidding too high, though, potential bidders may feel that there is no chance of that super-cheap bargain, the holy grail of online auctions, and so will not bid. Put the starting price much lower, and those max bidders will still put in a bid of over £100, but have their fingers crossed it won't get that high.

There's also the fact that you don't initially know how many other bidders there are, what type of bidder they may be and how much they are prepared to bid and under what circumstances. So how can game theory help here?

Pick and Mix

In the Prisoner's Dilemma, the best worst-case scenario comes from a strategy called, accurately, play safe. There are types of game where playing safe won't work though. Imagine you are playing a simple betting game with your friend. You each have a coin and can choose whether to put it on heads or tails. You then reveal your coins. If they match, you win. If they don't, you lose. You pay each other in biscuits according to this table:

		Friend plays:	
		Heads	Tails
You play:	Heads	1	-3
	Tails	-4	6

For instance, if you both played heads, your friend would give you one biscuit. If you played heads and your friend played tails, you'd give them three biscuits. This is called a zero-sum game: your winnings are the same as your friend's losses. When you have this kind of game, it doesn't make sense to play it safe: in this case, your best worst-case is to play heads as you stand to lose three biscuits, rather than four biscuits if you play tails. Your friends best worst-case is to play heads too, as they lose only one biscuit rather than six. So following this rule would lead to your friend always losing a biscuit every game. A dull game, and your friend would have to be phenomenally stupid to play it. Clearly, it would benefit your friend to play tails sometimes. But how often? And if they employ a mixed strategy, what should yours be?

Let's say there is an optimum proportion of the time you should play heads, and call it p. Then you should play tails 1 - p, the rest of the time. If your friend plays heads, we can write down what we expect you to win. You'll win a biscuit when you play heads, which is a pth of the time. You'll lose four biscuits when you play tails, which is a $(1 - p)^{th}$ of the time. I can write your expected winnings from this strategy as:

$$\text{Expected winnings} = 1 \times p - 4 \times (1 - p)$$

By multiplying out the bracket, remembering that a negative multiplied by a negative gives a positive, and writing it in a more algebraic fashion I get:

$$\text{Expected winnings} = p - 4 + 4p$$

A bit of simplifying:

$$\text{Expected winnings} = 5p - 4$$

This tells us that as p increases, so our winnings will increase. We can also make a similar expression for when your friend plays tails:

$$\text{Expected winnings} = -3 \times p + 6 \times (1 - p)$$
$$\text{Expected winnings} = -3p + 6 - 6p$$
$$\text{Expected winnings} = -9p + 6$$

This one tells us that our winnings decrease as p increases. Somewhere in the middle there has to be a value for p that gives us the best outcome. I can find this by equating the two expressions and solving the resulting equation to find p:

$$5p - 4 = -9p + 6$$

Adding 9p and 4 to both sides gives:

$$14p = 10$$

This means that:

$$p = \frac{10}{14}$$

So, you should play heads ten times in fourteen, or about 71 per cent of the time.

If I go through the same process from your friend's point of view, with them playing heads a qth of the time, I find that their winnings when you play heads are -4q + 3 and 10q - 6 when you play tails. Equating these:

$$-4q + 3 = 10q - 6$$
$$9 = 14q$$
$$q = \frac{9}{14}$$

So your friend should play heads nine times in fourteen, or about 64 per cent of the time. If you follow these strategies, who should be up on biscuits in the long run, though? Well, we have expressions for the winnings. Substituting p as ten fourteenths into either of your expressions gives:

$$5 \times \frac{10}{14} - 4 = -\frac{3}{7}$$
$$-9 \times \frac{10}{14} + 6 = -\frac{3}{7}$$

Oh, that's not good. The game, as you have it set up, means that in the long run you'll lose an average of three sevenths of a biscuit per game. Let's look at your opponent:

$$10 \times \frac{9}{14} - 6 = \frac{3}{7}$$
$$-4 \times \frac{9}{14} + 3 = \frac{3}{7}$$

Well, that makes sense – if you are losing three sevenths of a biscuit per game on average, they must be going to your friend. So maybe this game isn't for you after all.

Auction Action

There are several different types of auction. The traditional kind involves an auctioneer taking bids from people sitting together in a large room, often waving some kind of sign to place a bid. This is known as an English auction, and the highest bidder wins and pays the price they bid. You may also be familiar with sealed-bid auctions, often used by estate agents in the final stages of agreeing a price on a property – bidders submit their highest bid without knowledge of the other offers. Again, the winner pays the price they bid. EBay is not like this. In fact, it is a type of auction known as a Vickrey auction, named after Canadian economist and Nobel Prize winner William Vickrey (1914– 1996). A Vickrey auction is a sealed-bid auction, but the winner pays whatever the runner-up bid. This encourages bidders to bid their maximum bid, but with the incentive that they may end up paying less than this.

EBay follows this system, more or less, except that the current second-highest bid is public knowledge and multiple bids are allowed.

Sold!

So how can all this knowledge be put together to help you win online auctions? Well, you can use a mixed strategy, as per the biscuit game, but tailor it to the behaviour of the other bidders in the game. If you look at an item's listing, you can see the current winning bid (although not necessarily the highest bid), the number of bids and the number of bidders . You can use this information to decide on the nature of your fellow bidders.

If there are many bids from a few bidders, that could

Going Ape

In the heads or tails game mentioned above, playing a mixed strategy was key to success. A simplified version of the game – where winning gets you both coins and losing means your friend takes them both – would require you to use a strategy where you randomly chose heads or tails 50 per cent of the time. Sounds easy, right? Well it is, if you play it with another human.

If you play it with a chimpanzee, however, you are liable to come off second best. In 2014 a study from Caltech University in the USA showed that the apes were better at playing randomly than humans, resulting in a higher proportion of wins (they were rewarded with fruit treats rather than cold hard cash). This could be due to chimpanzees' excellent short-term memory, their more competitive society, or just simply that they are cleverer than us.

Some football teams use a similar game theory approach for deciding where to hit penalty shots in shootouts, but I don't think any have yet recruited a primate penalty analyst.

imply you are dealing with incremental bidders. The best strategy for these people is sniping – putting in your maximum bid in the last seconds of the auction, which denies them the opportunity to bid past you.

If there are few or no bids, this may imply that you are dealing with a bunch of snipers. The best way to deal with them is to get your maximum bid in early, as you will take precedence in the event that their last-second bids match yours.

If the current bid price is near what you consider to be the item's value without many bids having occurred, you are probably dealing with max bidders. In this circumstance they may have already beaten you to the punch, but max bidding or sniping will still work if you value the item more than them.

Notice that in all these situations there are none where incremental bidding is a good strategy. It may make the auction more fun for the bidder, but it won't help you to get a good price for the item. So whether you choose to be a max bidder or a sniper, good luck and happy bidding!

PART 5

And Relax

Work is done and the shopping is delivered. Time to chill, but not without understanding the mathematics behind how computers, streaming services and social-media stardom work.

CHAPTER 13

Unnatural Numbers

Relaxation, for me, often involves technology especially at the end of a hard day. I might watch a movie, flick through my social media or turn on some music. All these things rely on a simple piece of electronics called a transistor. A transistor is effectively a tiny switch without any moving parts, and a computer's microchips are packed with billions of them: it's what the computer uses to count, store data and perform maths and logic. No transistors, no modern computers.

Transistance is Futile

Transistors are tiny. The first iPhone used about 2 billion of them, whereas the latest uses more like 12 billion. Transistors are made from materials called semiconductors. As the name suggests, these materials are somewhere between conductors, which conduct electricity well, and insulators, which don't. They can be made to change between these two states, which is what makes them able to act as switches.

Transistors were developed alongside quantum mechanics, that murky area of physics that tries to explain how things work at the smallest scales. And quantum physics hinges on a very strange mathematical idea.

Summing Up

Mathematicians call the numbers you use to count the natural numbers: i.e. 1, 2, 3, etc. The natural numbers are a sequence of numbers that carries on going up forever. A series is what you call a sequence added together:

Sequence: 1, 2, 3, 4, 5, ...
Series: $1 + 2 + 3 + 4 + 5 + ...$

You probably associate quantum physics with a certain amount of weirdness. You may have heard of Schrödinger's Cat: the thought experiment where a quantum physicist's cat is shut in a box with a deadly poison that gets released after a random amount of time, and – according to quantum physics – the cat is both alive and dead in the box until someone checks on it, at which point the cat's actual status is revealed. Well, there is also some mathematical weirdness, and that weirdness states that, if I continue adding up the series of natural numbers all the way to infinity, the total would be slightly less than zero:

$$1 + 2 + 3 + 4 + 5 + \cdots + \infty = -\frac{1}{12}$$

Somehow, when you add all the natural numbers together, you end up with less than any of them. This is counter-intuitive, to say the very least. Most people, mathematician or not, feel that there must be something wrong. The various proofs of this sum, known as the Ramanujan Summation, use a bit of mathematical handwaving that upsets many mathematicians. But the proof of the pudding is that transistors work and our best theories use the Ramanujan Summation as part of the physics behind them.

Grandi Designs

The 'proof' requires us to look at two other series. The first, known as the Grandi Series after the Italian monk who studied it, goes like this:

$$G = 1 - 1 + 1 - 1 + 1 - 1 + ...$$

It continues forever, adding one and then subtracting one. Its value depends on how far I go, but it is always either 1 or 0. When I get to infinity (which I can't actually do, but anyway), would I be on 1 or 0? Mathematicians say that you can't know and so the series has no total. But there's a bit of a hack I can use which gets the maths professors upset. If I take the series away from 1, I get:

$$1 - G = 1 - (1 - 1 + 1 - 1 + 1 - 1 + ...)$$

If I expand the brackets I get:

$$1 - G = 1 - 1 + 1 - 1 + 1 - 1 + ...$$

The right-hand side looks familiar – it's G:

$$1 - G = G$$

Adding G to both sides gives:

$$1 = 2G$$

This means that G = ½, which kind of feels right because we know it would be either 1 or 0 and a half is right in the middle of these, but then again feels wrong because a half isn't actually either 1 or 0.

New Series

The second series I need to look at goes 1 - 2 + 3 - 4 + 5 - .... Again, the sum of this series varies as you go along, going from negative to positive, but I can do some more jiggery-pokery to get an answer if I keep on going to infinity. The series doesn't have a special name, so I'll just call it S. I'm going to add S to itself:

$$S = 1 - 2 + 3 - 4 + 5 - ...$$
$$S + S = (1 - 2 + 3 - 4 + 5 - ...) + (1 - 2 + 3 - 4 + 5 - ...)$$
$$2S = 1 + 1 - 2 - 2 + 3 + 3 - 4 - 4 + 5 + 5 - ...$$

Now I'm going to combine these selectively:

$$2S = 1 + (1 - 2) + (-2 + 3) + (3 - 4) + (-4 + 5) + ...$$

If I total each bracket, some magic happens:

$$2S = 1 + (-1) + (1) + (-1) + (1) + ...$$

Dealing with the brackets gives us an old friend:

$$2S = 1 - 1 + 1 - 1 + 1 - ...$$

That old friend being Grandi's Series, G, which we know has a total of a half:

$$2S = \frac{1}{2}$$

This Sucks

The sum of the Ramanujan Series makes most mathematicians very uncomfortable. How can the sum of a bunch of positive numbers ever be negative?

In 1948, Dutch physicist Hendrik Casimir (1909–2000) predicted that there should be a force between two metal plates in a vacuum, according to the rules of quantum physics, which require the strange result of the Ramanujan Summation. The existence of the Casimir force was finally verified in 1997, which in turn lends validity to the sum of all natural numbers being slightly less than zero.

Divide both sides by two and I can find S:

$$S = \frac{1}{4}$$

Again, another result I would never actually get if I mechanically added up the terms in the series, but we'll press on.

Series Finale

If I call the Ramanujan Series R and subtract S I get:

$$R - S = (1 + 2 + 3 + 4 + 5 + ...) - (1 - 2 + 3 - 4 + 5 - ...)$$

Some more dealing with brackets and rearranging:

$$R - S = 1 - 1 + 2 + 2 + 3 - 3 + 4 + 4 + 5 - 5 + ...$$

I can see that all the odd numbers will cancel out, and I'll have double all the evens:

$$R - S = 4 + 8 + 12 + 16 + \ldots$$

This is the Ramanujan Series multiplied by 4. Remembering that S is a quarter gives:

$$R - \frac{1}{4} = 4R$$

This means 3R must be minus a quarter:

$$-\frac{1}{4} = 3R$$

Dividing both sides by 3:

$$-\frac{1}{12} = R$$

There you have it, R, the Ramanujan Series, the sum of all the natural numbers to infinity, is slightly less than zero. And because this is true, transistors work and I can enjoy all the benefits of modern computing and smart devices, not to mention lasers, low-wattage LED lighting and MRI scans.

CHAPTER 14

Good Vibrations

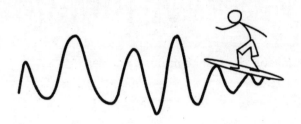

Now that we know that modern electrical devices depend on the mathematical voodoo of the Ramanujan Series, we can look at some more down-to-earth mathematical techniques that let us use the electronics for communication and entertainment.

Smile and Wave

Whether you are singing in the shower, calling a friend or listening to a podcast, you are experiencing tiny variations in air pressure on your eardrums which your brain translates into sound. It all depends on vibration – that is, things moving back and forward, whether it's the air particles, your eardrum or the speaker on your hi-fi. Vibrations can happen slowly or quickly, giving us low and high pitches respectively. You can also have larger and smaller vibrations – regardless of speed – which give us louder or softer sounds. The pitch (or frequency) of the vibration along with its size (amplitude) mean that we can represent sound as wave-like form.

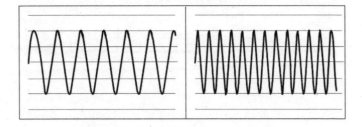

These two waves have the same amplitude (and so are equally loud), but different frequencies. In fact, the one on the right has twice the frequency of the one on the left – it goes up and down twice for every time the left hand one goes up and down once. If you put either of these waves through a speaker, you'd hear a very pure sound at a specific pitch. Sounds like these – called a sine wave – don't occur naturally very often, for reasons we'll see later. When you double the frequency, the pitch changes by an octave. The note that an orchestra tunes to at the beginning of a recital is an A where the vibrations occur 440 times per second. Double this to 880 and you get another A, the octave above the first. These notes sound similar – like a man and a woman singing the same song in tune, but with the woman's voice higher than the man's. Frequencies and pitches are measured in hertz – Hz – after the German physicist Heinrich Hertz (1857–1894). A hertz simply means 'per second' – something vibrating at 10 Hz is going backward and forward 10 times per second.

When we produce sound by speaking, or singing, or playing an instrument, the sound waves are much more complex than the sine waves we saw above. This is because of something called harmonics. Let's take the string of a guitar as an example. When you pluck it, the string wobbles back and forward and produces the pitch.

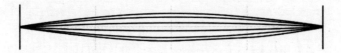

It takes a moment for the string to start wobbling as you pluck it, and then the volume lessens as time goes by.

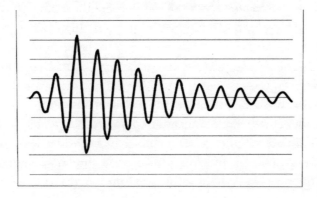

This still looks like a sine wave that gets louder and then softer, but a guitar doesn't sound like a sine wave. Why? When you pluck the string, it vibrates in more than one way. For instance, it can also vibrate like this:

These vibrations occur as the other ones do, but with twice the frequency, so they sound like the octave above, just like the two waveforms shown at the beginning of the chapter. The two waveforms get combined to give the overall sound of the plucked string.

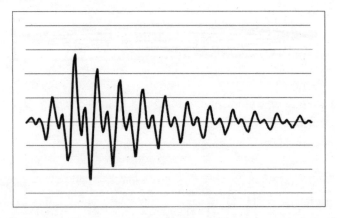

The string has many other modes of vibration, which all make the waveform more complex. We hear this as the timbre of the instrument. The varying ways that different instruments emphasize or de-emphasize certain harmonics lends them their signature sound.

Free Sample

With this understanding of how sound works, we can now look at how sound gets recorded for movies, podcasts or phone calls. To make an electronic recording of a sound, it has to be sampled. If you have a recording contract and are making an album at a recording studio, the audio will be sampled 44,100 Hz – CD quality. Where does this number come from? Well, they pulled it out your ear, actually.

When sound is sampled, it means that a snapshot is taken of the waveform at that time. The more snapshots, the more the sampled waveform will resemble the original waveform. It's the same with seeing things – if I show you two photos from a football match you get an idea what was going on at two separate moments, but if I take more and more photos you can follow the action. If I take twenty-four photos per second and you view them at that rate, you'd be watching a real-time movie of the match.

When you sample something in this way, you need to bear in mind something called the Nyquist–Shannon Theorem, named after two electronic engineers who did pioneering work on sampling in the 1920s and 1930s, just as broadcast radio was becoming a big thing. Nyquist and Shannon worked out that if you want to represent something well with sampling, you need to sample it at twice the rate of the thing you are sampling. Human ears have a range from about 20 Hz – the super-low notes on a tuba or church organ – up to 20,000 Hz, which would be an extremely shrill whistle or the high-pitched noise of an old non-flatscreen television turning on. Double 20,000 Hz and you see that you need to sample at least 40,000 Hz. The remaining 4,100 Hz are there to make the audio sit nicely with video for television. It's a similar story with video – twelve pictures per second is enough to convince your brain that you are watching a video, but we have to sample twice as fast to make it look smooth and fluid. Hence cinema shows films at 24 Hz, with television 25 or 30 Hz, depending on where you live. Intense action or driving computer games can be even higher.

Data Bundle

So, you've recorded your three-minute pop song. It has been sampled at 44,100 Hz, giving a grand total of 44,100

× 180 = 7,938,000 samples. Each sample contains 16 binary digits or bits. Binary is a number system that computers use, as opposed to the decimal system that we count in. A 16-digit binary number goes up to 2^{16} = 65,536. Each sample contains a number from 0 to 65,356 to describe the waveform at that point. So the song will be made up of 7,938,000 × 16 = 127,008,000 bits of information. To put this into more familiar territory, a byte is 8 bits, so 127,008,000 ÷ 8 = 15,876,000 bytes, or roughly 16 megabytes.

Quelle Surprise

The iPod, perhaps the most famous portable media player but not the first of these, was launched in 2001, six years before the first iPhone. It was a portable device that contained a hard drive where you stored the music. This was a big deal, as previously we had been wandering around with tapes and compact discs, which took up a lot of physical space. The first iPod had a 5-gigabyte hard drive, which meant I could upload about 400 songs like our example. This was great, but was there a way to get more on there?

In the mid-1990s, people had begun to share files over the internet and it wasn't long before music and video became part of this. Broadband internet was not a thing yet, so sharing even our 16 MB song would take a while. Dial-up internet, which used the existing phone lines, might manage to transfer information at a rate of 50 kilobits per second, which meant downloading our song would take 16,000,000 ÷ 50,000 = 320 seconds – over five minutes.

Naturally, computer scientists worked on ways to reduce the size – or compress – these music files. One was called

Phenomenal

Joseph Fourier (1768–1830) had a very interesting life. Orphaned at an early age, he was initially training to become a Benedictine priest, before he caught the mathematics bug and became a teacher and active revolutionary during the French Revolution. He was a scientific adviser to Napoleon Bonaparte when he attempted to annexe Egypt in 1798, but returned to France in 1801 after the French surrendered to the British forces. There, he split his time between being the governor of the area around the city of Grenoble and working on various mathematical pursuits involving heat transfer and the maths discussed below. He was also the first to understand the greenhouse effect – that the Earth's atmosphere traps heat from the Sun, keeping it warmer than the energy falling on it from the Sun would suggest.

Fourier said about his various studies: 'Mathematics compares the most diverse phenomena and discovers the secret analogies that unite them.' This, I think, gives an insight into one of the many reasons that people like me love mathematics.

Mp3, but it relied on the mathematics of a Frenchman, Joseph Fourier (1768–1830) to get the job done.

Fourier worked out that any equation, even the most complex, can be approximated by adding together simpler equations. It's the reverse of what we did above, looking at the guitar strings. We added together two simple waveforms to get a more complex one. Fourier found a way to do this in reverse. If you think about it, Fourier's method does exactly what your ears and brain do when you listen to music. The sound from all the different instruments and voices comes as one mixed signal to your ears. Your brain is able to distinguish these as guitars, drums, voices, etc.

The inventors of Mp3 used Fourier analysis to split the music signal into thirty-two frequency bands. Firstly, any frequency bands beyond the range of human hearing can be discarded. Secondly, any frequency bands with little or no signal in them can also be discarded without significantly reducing the perceived quality of the song.

There is also the fact that the human ear is better at hearing certain frequencies than others within our hearing range. We are best at hearing sounds in the region from 2 to 4 kHz, which coincides with the spoken voice. Sounds that are outside of these frequencies sound quieter to us, even when they are played at the same volume. This means that a frequency band with a quiet signal could also be discarded without the listener noticing.

Put all these factors together, and the Mp3 algorithm can reduce the amount of raw data used to store the song. A good quality Mp3 uses 128 kilobits each second. Our songs would then contain $128{,}000 \times 60 \times 3 = 23{,}040{,}000$ bits, which is $23{,}040{,}000 \div 8 = 2{,}880{,}000$ bytes, or 2.88 MB, less than a quarter of the CD file. Now that we tend to store our music digitally, this means they take up less space on hard drives or cloud storage, as well as making them easier to stream over wi-fi and data connections.

Hit the High Notes

As a child, I used to get the bus to school. I lived in London, so I used the famous red double-deckers. I noticed that sometimes, when the bus was stopped at lights or a stop, the vibration of the engine would make the poles shake and cause an awful racket. It wasn't until I was at university that I discovered that this phenomenon is called resonance. We looked briefly at the resonance of singing in the shower in Chapter 2.

All objects have a frequency at which they resonate, which means there are certain frequencies where inputting energy produces the most vibration. If you think of pushing a child on a swing, there are certain times when pushing them will increase their speed and times when it won't. The notes that sound best when you sing are due to the resonant frequencies of your chest, mouth and head, and the air spaces within them. On the bus, the frequency of the engine hit the resonant frequency of the poles, making them rattle in their housings.

We've all heard the story of opera singers being able to shatter glass with the power of their voices. How could this be possible? Well, wine glasses are well known for their resonance – every time someone tings a glass with their cutlery before a speech shows that a glass, with a nice cavity for producing sound, resonates at a particular note. If you can reproduce that note with your voice, the vibration will pass through the air and make the glass vibrate. If it vibrates enough that the microscopic flaws in the glass break, then you've made it. To be in with a shout, you need a powerful voice.

CHAPTER 15

Reach for the Stars

For those of us that remember a time before smartphones, the rise of social-media platforms has been prodigious. Not only are they a great way to stay in touch with your extended network of family, friends and acquaintances, but they also allow you to stay in touch with your favourite brands, artists and celebrities. You can reach out to people you admire and – if you are fortunate – they will respond.

So celebrities can be contacted at the click of a button – or can they? You are actually more likely to hit a social-media manager, posting on their behalf. Much better if you can get someone they personally know and trust to pass on (or forward, share, retweet, pin, etc.) your message. But how do you get to their inner circle?

Six Degrees

Mathematicians, economists and political scientists have been interested in social networks since well before the internet. Back when inventions like long-distance telephone lines and the increasing affordability of cars allowed people to have more far-flung contacts, enquiring minds wanted to know how interconnected people were and how this was changing.

We can model this by making a few assumptions. Let's say the typical person has fifty friends and acquaintances close enough that they could ask for a favour. This means that, if you take any two people who know each other, they have a pool of 50 × 50 = 2,500 acquaintances between them:

Would you Adam and Eve it?

Every cell in your body contains genetic information that is half from your mother and half from your father. Within your cells are mitochondria and these have their own DNA which you inherit exclusively from your mother. Analysing this mitochondrial DNA means that scientists can look for Mitochondrial Eve – the woman that all living humans are related to. Current understanding suggests that this woman was alive one hundred to two hundred thousand years ago. Y-Chromosomal Adam, studied by looking at the Y chromosome that is passed down the male line, is dated to two hundred to three hundred thousand years ago. This makes it highly unlikely that Y-Chromosomal Adam and Mitochondrial Eve were a couple.

the first person's fifty friends each have fifty other friends. This assumes that they don't have any friends in common, which is unrealistic, perhaps, but let's stick with it for now. Once we get up to three friends, they have 125,000 in their group. As I add friends the numbers go up by a factor of fifty each time. When I reach six friends, their social network extends to over 15 billion people. There aren't 15 billion people in the whole world. The upshot is, with this model, I could get in touch with anyone through a chain of about six friends and acquaintances. This result is called six degrees of separation and serves to demonstrate how 'small' the world is.

This is a model though, and while I kept the number of acquaintances low, obviously some of them will know each other and reduce the overall number. How could we work out something more realistic?

Pen Pals

In the 1960s, American psychologist Stanley Milgram (1933–1984) wanted to answer this question, so he devised a simple experiment. He put an advert in newspapers in cities far away from Boston, USA, calling for volunteers who felt they were well connected. He sent letters to the respondees giving the name of a person who lived in Boston, with instructions to either forward the letter directly to that person, if they knew him or, if they didn't know him, to forward the instructions to someone who they thought might know him.

With the data he gained, Milgram concluded that most Americans were separated by about six people at most, corroborating the result from our experiment above.

Bacon

There were, of course some problems with Milgram's experiment: many people received the letter and did not follow up on it, which means that the longer the chain, the less likely it was to ever get to the target. If you limit the population you are dealing with, it becomes easier to avoid this problem.

Enter Kevin Bacon, the US actor who has been in over sixty movies in his long career. The Kevin Bacon game was invented by some students in the 1990s: the aim of the game is to choose another actor and link them back to Kevin Bacon via their co-stars. For example, if I take Saoirse Ronan: she was in *Atonement* with James McAvoy, and James McAvoy was in *X-Men: First Class* with Kevin Bacon. Two steps. This gives Ronan a Bacon number of 2 and McAvoy a Bacon number of 1. Kevin Bacon himself, of course, has a Bacon number of zero.

Mathematicians have their own – admittedly much more nerdy – version of this game. Your Erdős number is how many mathematicians are required to link you to the Hungarian mathematician Paul Erdős, a prolific collaborator and co-author of mathematical papers (including the coffee guys from Chapter 1). For instance, Albert Einstein has an Erdős number of 2 – he published a paper with someone who published a paper with Erdős. Some unlikely people have an Erdős number: actor Natalie Portman, who studied psychology at Harvard University, published a paper with collaborators that leads to her having an Erdős number of 5.

If you combine these two games, you can play the super-niche Erdős–Bacon game, by adding Erdős and Bacon numbers. Portman has a Bacon number of 2, so her Erdős–

Just Following Orders

Stanley Milgram is also known for a less salubrious experiment. In 1961 he got volunteers to give electric shocks to people to study the effect of punishment on memory. As the experiment wore on, the voltage of the shocks would increase to dangerous levels.

Unbeknown to the volunteers, the actual experiment was to see how far they would go, despite the pleading, screams and eventual silence of the actor playing the victim. Despite their misgivings, over 60 per cent of people would administer the highest level of electricity.

Shocking.

Bacon number is 5 + 2 = 7. I'll leave you to figure out some others.

One Big Family

You can use the same maths as social networks to prove that all humans must share ancestors: that we are all, in fact, cousins. Every human has exactly two biological parents, who each have exactly two biological parents, and so on. Essentially, you need two parents, four grandparents, eight great-grandparents and so on up your family tree.

Let's go back a thousand years and assume that, on average, your ancestors reproduced every twenty-five years. That gives forty generations. So, to work out how many ancestors you need forty generations ago, I multiply by two forty times. $2^{40} = 1,099,511,627,776$. You need more than a trillion people in the fortieth generation behind you.

Sounds like a lot? It is, especially when you consider it is estimated that only around 100 billion people have ever lived.

What does this mean? Well, for one thing it means that a lot of your ancestors must appear more than once in your family tree. This is a nice way of saying that we are all inbred to some extent. It also means that many of your ancestors must also be my ancestors, so everyone is related, no matter how distantly.

Social Networking

Say you want to get noticed by your celebrity fan-crush by a chain of acquaintances. Your chances of it working depend on the size of your network. The typical Facebook user has about three hundred friends. The typical non-celeb Twitterer has about four hundred followers. The typical Instagrammer has about a hundred and fifty followers. But you can pretty much guarantee that you know someone, who knows someone, who knows someone, who knows someone, who knows someone, who knows the person you want to get in touch with.

PART 6

Bedtime

It's been a long day, but more maths awaits us: the mixing of the perfect bath, and the graphs that show how to get the best night's sleep.

CHAPTER 16

Sunrise to Sunset

The words sunrise and sunset give a false idea, the idea that the Sun is doing something at these times. It isn't: as we all know, it is the Earth spinning that makes it appear that the Sun comes up in the East and sets in the West. Depending on where you live in the world, the time of sunrise and sunset will vary according to the time of year. But why?

Northern Light

The Earth is more or less a sphere. The radius at the equator is about 30 km more than at the poles, which may seem like quite a big difference. The distance from the centre of the Earth to the equator is 6,380 km, so 30 km is less than half a per cent of this. We measure how far away from the equator we are in degrees of latitude. I live in York in the UK, which sits at 54° north of the equator. This means the angle between the equator and a line going through the centre of the Earth to York would make a 54° angle.

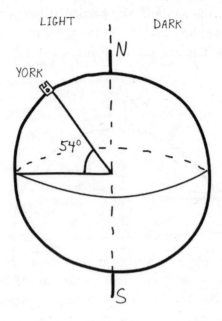

This is really quite far north. During the summer solstice, the daylight lasts for just under 17 hours. During the winter solstice we only get about 7 hours and 22 minutes. Quite a difference. The diagram above, where the equator is horizontal, represents an equinox. At this point, York, and everywhere else for that matter, gets 12 hours each of daylight and darkness: equinox comes from Latin words meaning equal and night.

If the Earth always presented itself like this to the Sun, we'd all have equal days and nights across the world. Scientists think that when the Earth was forming it took a few hits from mini-planets which knocked it off centre a bit. As a result, the Earth's axis of rotation is not vertical, but tilted at 23.4° to the vertical. As the Earth orbits the Sun over the course of a year, that tilt changes the intensity and duration of the sunlight any particular latitude receives, and gives us the seasons. On the 21st June, the summer solstice, the diagram looks like this:

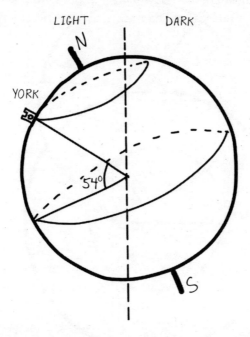

You can see that much more of York's motion around the world is in the daylight, so the day is longer. The situation reverses in winter, when most of York's motion is on the dark side of the Earth. The actual maths involved in working out how much daylight anywhere gets is a bit tricky, but not beyond school maths. But fair warning – you might want to skip the next bit if trigonometry is something you'd rather not reacquaint yourself with. What comes next is some of the trickiest maths in this book.

Tricky Trigonometry

Let's work out how many hours of sunlight York should get on the summer solstice.

Imagine I sawed the top of the Earth off at 54° latitude, a bit like taking the top off a boiled egg, and then we looked down from above the North Pole. I'm going to mark a few points on it and call it Diagram 1, as we'll refer back to it later:

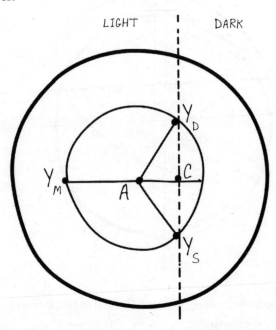

A is the centre of the circle, on the same line as the centre of the Earth and the North Pole. York at dawn, midday and sunset are marked respectively Y_D, Y_M and Y_S. To work out how long York is in the sunshine, I need to know what fraction of its motion is in the light, and to do that I need to work out the proportion out of 360°. To do this, I need to know the distances from Y_M to A and from A to C.

I know York is at a latitude of 54° and that the radius of the Earth is 6,371 km, so a side-on view would look like this:

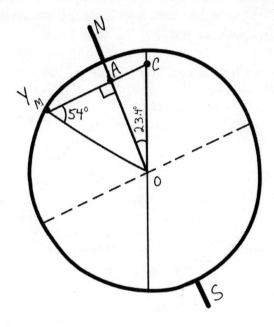

In triangle Y_MAO, where O is the centre of the Earth, I can use trigonometry to say:

$$\cos 54° = \frac{Y_M A}{OY_M}$$

I know OY_M is the radius of the Earth:

$$\cos 54° = \frac{Y_M A}{6,371}$$

Multiplying both sides by 6,371 gives:

$$6{,}371\cos54° = Y_MA$$

A quick tap on the calculator later, I have Y_MA = 3,745 km to the nearest kilometre. So now I know the circle around the Earth at 54° North has a radius of 3,745 km. I'm going to use this fact a bit later.

To work out the distance from A to C, I need to know the distance from A to O. In triangle Y_MAO, where O is the centre of the Earth, I can use trigonometry to say:

$$\sin54° = \frac{AO}{OY_M}$$

I know OYM is the radius of the Earth:

$$\sin54° = \frac{AO}{6{,}371}$$

Multiplying both sides by 6,371 gives:

$$6{,}371\sin54° = AO$$

Another quick tap on the calculator gives AO = 5,154 km to the nearest kilometre. Now I can look at triangle ACO. I know that the angle at O is 23.4°, so I can say:

$$\tan23.4° = \frac{AC}{AO}$$

I've just worked out AO, so I can use similar working to the above to get:

$$5{,}154\tan23.4° = AC$$

This gives AC = 2,230 km to the nearest kilometre. Going back to Diagram 1, I can now look at triangle ACY_S and say:

$$\cos A = \frac{AC}{AY_S}$$

I know AC and AY_S is the same as AY_M that I worked out first of all, so:

$$\cos A = \frac{2230}{3,745}$$

This means that:

$$A = \cos^{-1}\left(\frac{2230}{3,745}\right)$$

So the angle at A is 53.5°. Triangle ACY_D is the same as triangle ACY_S, which means the angle of York's motion that is in the dark is 53.5 + 53.5 = 107°, which leaves 360 - 107 = 253° in the daylight. We can then work out this as a proportion of 360° and multiply by 24 hours to get the hours of daylight for York, or indeed anywhere on that latitude, on the summer solstice:

$$\frac{253}{360} \times 24 = 16.87 \text{ hours}$$

This is 16 hours and 52 minutes of potential daylight. Over the course of the year, as the Earth moves around the sun, varying amounts of the 23.4° tilt make higher latitudes have longer or shorter days.

Natural Light

The Sun is a gargantuan nuclear furnace over a million times the size of Earth. Ninety-eight per cent of the energy it emits is in the form of photons – massless particles that form the electromagnetic spectrum:

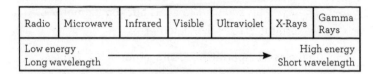

Radio	Microwave	Infrared	Visible	Ultraviolet	X-Rays	Gamma Rays
Low energy						High energy
Long wavelength		⟶				Short wavelength

The energy and wavelength of the photon determines its behaviour and what we call it. Longer wavelength photons have less energy: we call these radio waves. Shorter wavelengths bring us to microwaves – exactly what we use to heat food in microwave ovens, but also for wi-fi and radar. We then hit infrared, which we feel as heat. Then comes a tiny slice of the spectrum we call visible light – all the colours of the rainbow. After that, we reach ultraviolet – the stuff that causes sunburn and skin cancer. Beyond that, we move into X-rays, which have enough energy to pass through all but the boniest parts of our bodies. Finally, gamma rays, which will pass right through you.

The Sun emits all wavelengths of visible light. When you combine all the wavelengths of visible light, you get white light, so the Sun is white. But this is not our perception of it a lot of the time. If you look at art, whether its Van Gogh or your child's drawing stuck on the fridge, the Sun is invariably shown as being yellow or orange. Why is this?

If you went up to space (see Chapter 6), the Sun would appear white. Down on Earth, the atmosphere affects the way we perceive it. At school, you would have seen

Move Closer

The orbit of the Earth around the Sun is almost, but not quite, circular. This means that the Earth is 5 million kilometres closer to the Sun at its closest point than at its furthest. This sounds like a lot, but the diameter of Earth's orbit is 300 million kilometres.

The closest approach occurs around 3rd January, ironically in the middle of winter for the northern hemisphere. Furthest approach is six months later, on 3rd July. So, the southern summer is warmer than the northern for being closer to the Sun, but this balances out because more of the southern hemisphere is covered by water which absorbs most of the 7 per cent increase in solar energy received.

the experiment where white light is split up into the visible spectrum by a glass prism, or maybe you know it from the famous album cover for Pink Floyd's *The Dark Side of the Moon*. When light goes from one medium to another, it bends – hence why straws look broken when a glass is viewed from certain angles. But light doesn't bend evenly. The amount it bends depends on the wavelength of the light rays and so it splits up to make pretty rainbow patterns.

Exactly the same thing happens with the Earth's atmosphere, which itself is a mixture of different gases and vapours. The shorter wavelengths, at the blue end of the spectrum, get scattered first, which is what makes the sky appear blue. The light that gets down to us is therefore shifted towards the redder end of the spectrum. The more atmosphere in the way, the more the light bends, so at sunrise and sunset, when the light is coming in obliquely through lots of atmosphere, we get lots of orangey reds coming through. It works with the Moon, too – the lower in the sky, the redder the Moon appears.

We have evolved for this, and our eyes and brain use the colour of light we receive to tell us how tired we should be (see Chapter 18).

CHAPTER 17

Bath and Bed

In Chapter 2 we looked at the mathematics of the shower which many of us use to get ourselves refreshed and awake for the day ahead. When getting clean quickly is not your main priority, many people enjoy the luxury of a nice hot bath. But why do we find hot water so relaxing?

Buoyed Up

By the evening, your muscles have been working hard all day. Even, or maybe especially, if you've been sat at your desk, your muscles have been holding you in position and are tired. They have to work against the relentless force of gravity.

When an object floats or is submerged in water, the object has to push water out of the way. This is why the level of the bath rises when you get in it and what helped Archimedes to develop the principle named after him: an object immersed or floating in liquid displaces its own weight in liquid. Not only that, but the liquid pushes back with a buoyancy force equal to the weight of liquid displaced.

Let's take an example. Imagine I dropped an iron cannonball into the ocean. Cast iron has a density of about 7.2 g/cm³, whereas seawater's is 1.024 g/cm³. If the volume of the ball is 4,000 cm³, the ball would have a mass of 4,000 × 7.2 = 28,800 g or 28.8 kg. The same volume of water would have a mass of 4,000 × 1.024 = 4,096 g or 4.096 kg.

We need to compare the weights (rather than the masses) of the ball and the water, using W = mg:

$$W_{cannonball} = 28.8 \times 9.8 = 282 \text{ N}$$
$$W_{water} = 4.096 \times 9.8 = 40 \text{ N}$$

So, the cannonball has a force of 282 N pulling it downwards and a buoyancy force of 40 N pushing it upwards. As expected, the weight wins and the cannonball will accelerate towards Davy Jones' Locker.

The density of a human is very similar to that of water. It varies a little bit due to body composition: fat is less dense than muscle, so the more fat you have, the better you will float. Essentially, people just about float in water and so the force of gravity is cancelled out by the buoyancy from the water. Result: your hard-working muscles get a well-deserved chance to rest.

In Hot Water

In terms of mathematics, heat transfer is a tricky business, but there are some basic rules of thermodynamics we can employ in the bath situation. Heat travels from a hot object into its surroundings until everything has the same temperature. The amount of energy that gets transferred follows this equation:

$$\text{Energy} = mc\Delta T$$

In this equation, m stands for mass and ΔT means the change in temperature. The c – specific heat capacity – requires a little more explanation. It is a little like density: it's a measure of how much energy is required to heat a kilogram of a substance by one degree Celsius. Water has a specific heat capacity of 4,184 Joules per kilogram for every degree Celsius. A good hot bath is about 45 °C, but water comes out the hot tap at 55 °C and the cold at 7 °C. Is there a mathematical way of working out how much of each I'll need?

The key is to realize that the energy from the hot water will flow into the cold water: we want enough hot water to heat the cold water up to 45 °C and enough cold water to cool the hot water down to 45 °C.

The energy required to heat one kilogram of the cold water up by 38 °C to my target temperature of 45 °C is given by:

$$\text{Energy} = 1 \times 4{,}184 \times 38$$
$$= 158{,}991 \text{ J}$$

The energy that one kilogram of the 55 °C hot water needs to transfer to drop to 45 °C is:

$$\text{Energy} = 1 \times 4{,}184 \times 10$$
$$= 41{,}840 \text{ J}$$

I now divide the two numbers:

$$158{,}991 \div 41{,}840 = 3.8 \text{ (one decimal place)}$$

This tells me that every kilo of cold water needs 3.8 kilos of hot water to warm it up. But how much water do I need in total? A normal bath is usually about 1.5 metres long, 80 centimetres wide and about 45 centimetres deep.

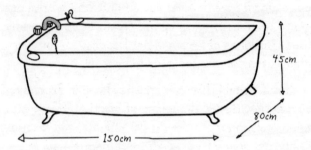

If I treat the bath as a cuboid and work in centimetres, half-filling it to a depth of 22.5 cm, this gives the water a volume of:

$$\text{Volume of cuboid} = \text{Length} \times \text{Width} \times \text{Height}$$
$$= 150 \times 80 \times 22.5$$
$$= 270{,}000 \text{ cm}^3$$

Water has a convenient density of 1 g/cm^3, so this water has a mass of 270,000 g or 270 kilograms. To split this up into water from each tap, I need to spot that, if every kilo of cold needs 3.8 kilos of hot, I need 4.8 kilos of hot and cold water to get a bit of water at the right temperature.

$$270 \div 4.8 = 56.25$$

So I will need 56.25 kg of cold water and 56.25 × 3.8 = 213.75 kg of hot water. This equates to 56.25 litres of cold and 213.75 litres of hot to make the perfect bath.

Ice-Cream Diet

When you eat something cold, like ice cream, heat from your body has to warm it up. Someone worked out eating ice cream gives you around 2.5 Calories per gram, whereas heating up ice cream in your tummy uses 17 calories per gram, a net loss of 14.5 for every gram eaten. Sounds too good to be true!

It is. Capital C Calories are an old way of expressing kilocalories, that is thousands of calories. This makes the calculation more like 2,500 - 17 = 2,483 calories gained per gram of ice cream consumed.

In the UK, all food is now labelled with kcal to avoid this confusion.

Pump it Up

Although human body temperature varies over the course of the day (see Chapter 18), it averages out at around 37 °C. But your temperature isn't the same everywhere. Further away from your core, extremities such as your hands and feet can be several degrees cooler – hence the invention of socks and gloves to keep these bits warm in cold weather.

When you slip into a hot bath, your body responds to keep your core temperature from increasing. It makes the blood vessels in your extremities dilate to divert all the hot blood to these colder parts, hence the flush many of us experience on our skin from a bath, or even just being hot. This increases the volume of your blood vessels, effectively reducing your blood pressure. Your blood will pump faster to accommodate this, making lying in the bath actually equivalent to taking a walk, calorie-burning-wise.

The extra blood flow has other benefits too. Muscles feel tired due to the build-up of lactic acid, and blood helps to transport this away. Muscles and connective tissues such as tendons and ligaments become more elastic when warm, adding to the feeling of literal and figurative relaxation. Pain signals from nerves get confused with the heat signals, making them get lost in the noise, which means that baths can also be an effective painkiller.

Getting Steamy

As you enjoy your bath, the heat transfer continues. The hot water's thermal energy goes into you, the bath and the air around the bath, making the air steamy. As your core temperature rises, as well as flushing your skin with blood, your body begins to sweat.

Sweating cools you down as thermal energy from your body heats the sweat and causes it to evaporate. To make liquids turn into gases, there is an energy barrier known as latent heat – essentially, you have to give the molecules in the liquid enough energy to break free of the liquid and float off. As the molecules float away, they take the thermal energy with them.

Sweating has other benefits, too. As your heart rate is raised by the bath and you are getting warm, your body may think you are engaging in exercise and begin to release hormones associated with this. Dopamine, which is involved in pleasure and reward, and serotonin, which is involved with feelings of happiness and well-being, start to be secreted in your brain. Not only are you physically warm, but your brain makes you feel all warm and fuzzy, too.

Sleepy Time

As we'll see in the next chapter, body temperature plays a big role in sleep. When you get out of the bath, your temperature will begin to fall as you leave the hot water and steamy environs of the bathroom. A falling body temperature is one of the signals to your body that it's time to get ready to sleep and so a bath a little while before bedtime can jump start this process. In fact, water-based passive body heating (as scientists call it) or a hot bath or shower (as you or I would call it) before bed is a well-established method for helping people with sleep disorders such as insomnia.

When you add the natural break that a bath allows, the space to think or let your mind wander as you see fit, it's little wonder that a nice bath works wonders for a stressed body and mind. Everyone from the Romans to the Japanese swear by bathing in hot water. In England we even have a city called Bath!

CHAPTER 18

Waves of Sleep

There is nothing better than a good night's sleep. Yet this simple, natural process is startlingly complex and eludes many of us on a regular basis. In this chapter we'll look at the cycles that need to align to let you slip beneath the waves of sleep and the mathematics of what we should do – and not do – to make it happen.

Rhythms of the Night

Your body, and indeed that of most other plants and animals, has its own 24-hour clock. Known as the circadian rhythm, it is what tells your body when to sleep and when to wake up. It is a very clever system and can regulate your sleep-wake cycle whether you are hot or cold and whether you are in the long days of summer or the short ones of winter.

There are three main internal factors that govern and are governed by your circadian rhythm, a complex feedback loop that we still do not completely understand. The first is the secretion of a hormone called melatonin. Essentially, your brain makes this hormone when it is dark, and it tells your body that it is sleepy time. When it is light, your brain does not make any of this – wakey time. Newborn babies take a while to sort this reaction out, which is why it takes them some time to learn to sleep at night.

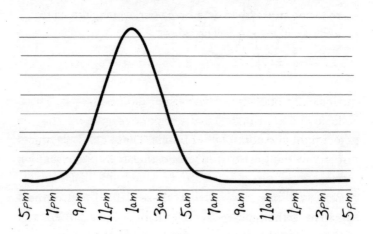

As you can see, melatonin starts to build up at around 8 p.m., peaks in the middle of the night and is back to a low level by around 7 a.m. How does your brain know when it is dark?

A Light in the Dark

Your body's light detectors are your eyes. The retina at the back of your eye contains three main light-sensing cells. Rod and cone cells allow us to see both colour and shape, depending on the level of light. The third kind was only discovered in the 1990s and have the less-catchy name of

Laughter is the Best Medicine

While this may not be true if you have a serious disease or injury, it is true for lowering your cortisol levels. Laughing releases endorphins – hormones which make you feel happy and are natural painkillers – as well as supressing cortisol production. It's a win-win for reducing stress, so maybe swap your evening drama for a sitcom if you are feeling tense.

intrinsically photosensitive retinal ganglion cells. These cells do not contribute directly to our eyesight, but they do detect light and control your pupils. These cells are linked directly to the part of your brain that runs the circadian clock.

Research has shown that these IPRGCs are most sensitive to the blue end of the visible spectrum. We saw in Chapter 16 that these wavelengths of light get scattered first and in the evenings the light is less blue, so your IPRGCs pick up less light and signal the brain that it is time to get ready for sleeping.

Unfortunately, most people nowadays prolong the available light with artificial ones. We light our streets and houses after dark as well as look at the screens of televisions, computers and smartphones. These all emit that blue shade of light that tells your brain that it isn't bedtime yet. Fortunately, many of these devices now have a night mode, where you can set the screen to reduce the blue wavelengths it emits so as not to confuse your body clock.

Hard Rocking

There are several well-documented examples of people staying awake for extended periods. In 1964, American seventeen-year-old Randy Gardner stayed awake for just over eleven days as part of a sleep-deprivation study, but not without experiencing slurred speech, attention deficit and hallucinations. This record was allegedly surpassed by Maureen Weston, a British woman who was going for the longest rocking-chair marathon. At fourteen-and-a-half days, she may hold the record, but the Guinness Book no longer lists it, in order not to encourage people to attempt to break this dangerous record.

You can also really confuse your circadian rhythm by travelling east or west a long way very quickly. Travelling long haul is tiring in itself, but when you arrive somewhere ready for bed but step out of the airport into the middle of a bright sunny day, it can completely throw out your circadian rhythm. Jet lag, as it is called, is generally worse if you travel east. This is because travelling east effectively shortens your day and it is hard for most people to get to sleep earlier than expected. Travelling west makes the day longer, meaning that, provided you can stay awake long enough, you can resync your circadian rhythm by staying up late. In that circumstance, the blue light from screens could actually be helpful, so consider watching the in-flight movies rather than nodding off in your chair.

Staying Cosy

The average core temperature for a human is about 37 °C. This varies a bit from person to person and day to day, but your temperature also varies throughout the day. A bit like a piece of machinery, your body needs to be warmed up when it is active during the day, but cooler at night when you are resting. Hence your temperature fluctuates like this:

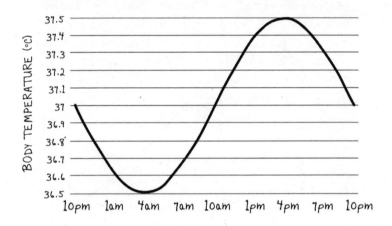

Our temperature increases from a low at around 4 a.m. to a high about twelve hours later. The changing temperature is part of the circadian rhythm, and the drop in temperature in the evening is a big sleep cue, which can be helped along by a bath or shower in the evening (see Chapter 17). The opposite effect works, too – a cold shower in the morning will provoke your body into warming itself, matching the increasing body temperature profile for the morning hours.

So, while no one likes getting into a cold bed, staying cool at night will allow your body temperature to do its thing.

Stress to Impress

The third factor that governs the circadian rhythm is the steroid hormone cortisol. Cortisol is often called the stress hormone – it is released along with adrenaline as part of the fight-or-flight reflex. It plays a role in your blood sugar levels, keeping you fuelled up for the day. It is also released during intense exercise.

Your levels of cortisol follow a natural pattern during the day – high in the morning to get you going and tapering off over the rest of the day:

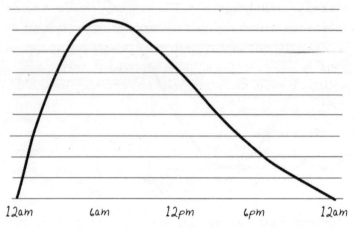

This works well for your typical hunter-gatherer. Cortisol levels start to increase from about midnight to reach effective levels as you are waking up. They peak at about 8 a.m. and then taper off. A stressful event, like a fight with a sabre-tooth tiger or hunting a mammoth, would give you a temporary spike, after which your levels would naturally return to normal for that time of day.

You probably don't follow a hunter-gatherer lifestyle and are not regularly in fight-or-flight mode. Instead, you probably experience some degree of stress as a symptom of modern life. Stress causes cortisol to be released too, which will interfere with your cortisol curve and can make it hard to relax and wind down for sleep, or perhaps have you waking early and not being able to get back to sleep.

To Sleep, Perchance to Dream

From what we've seen so far, it is clear that our bodies have evolved to enable us to spend the hours of darkness asleep. No one is entirely sure why. Theories vary from not being able to see well at night, or trying to conserve energy, or giving your brain downtime to make repairs and upgrades. What is known is that animals deprived of sleep don't do well; their immune system weakens and they eventually die.

Afterword

I hope you have enjoyed this book and that it has given you an understanding of how mathematics can help you negotiate life a bit more readily.

If you have enjoyed reading about cars, trains and rockets moving around, I recommend *What If?* by Randall Monroe, which takes things to extremes with great humour. You may also enjoy *An Equation for Every Occasion* by myself, which again takes things to extremes but focuses on how you can maths your way out of ridiculous problems.

If you have enjoyed the snippets about the mathematicians and scientists and their contributions, then you may enjoy *Cracking Mathematics* by Colin Beveridge, which is more accessible than most history of maths books. You could even try *From 0 to Infinity in 26 Centuries*, again by myself, which is a readable timeline of the story of maths.

Above all, I'd love it if you could help combat the spread of maths anxiety which is prevalent – and even fashionable – in this day and age. When children see or hear adults profess, even humorously, how bad they are at mathematics, it makes it seem okay for them to give up on the subject. It fosters a fixed mindset that implies that we are born able to do maths – or not – which is simply untrue.

So, if you are a parent of school-age children, the more involved you are with their education the less likely they are to fall prey to maths anxiety. If you learn alongside them, adopt a growth mindset and say things like 'I don't know how to do that, but maybe we can look it up together' rather than 'You're on your own, kid, I'm rubbish at maths,' then you'll put them in the best place to enjoy a compulsory but fascinating subject.

Maths matters.

Glossary

Acceleration
The change of speed in a given time

Algorithm
A list of instructions or mathematical operations that completes a task or solves a problem

Amplitude
The difference between the peak and the trough of a wave

Area
The amount of space a two-dimensional shape occupies

Axis (pl. axes)
A number line; two are often used at right-angles to create a coordinate grid to show graphs and shapes

Binary
A counting system that uses only zeroes and ones, used by computers and other digital machines

Bit
A binary digit

Buoyancy
The force that lifts objects submerged in a liquid

Byte
Eight bits

Calorie
A unit of energy, typically used for food and drink

Celsius
A unit of temperature based on the boiling and freezing points of water

Circumference
The distance around a circle; its perimeter

Cuboid
A solid with six rectangular faces set at right angles to each other

Density
An object's mass divided by its volume; an indication of how heavy a substance is

Diameter
The distance across a circle, through its centre

Dodecahedron
A solid with twelve faces

Equation
A mathematical expression containing an equals sign

Equinox
The time when the day and night are equal lengths

Expand
To multiply out a set of brackets

Expression
A set of mathematical symbols, numbers and letters

Formula
An equation or inequality that shows a mathematical or scientific law

Frequency
How many waves occur in a given amount of time

g
The acceleration due to gravity at the Earth's surface, equal to 9.81 m/s^2

Gravity
The force that causes masses to attract each other

Hypotenuse
The longest side of a right-angled triangle

Icosahedron
A solid with twenty faces

Inequality
A mathematical statement similar to an equation, but signifying a difference in value

Joule
A unit of energy

Litre
A unit of volume, usually used for liquids

Mass
A measure of the quantity of physical matter that makes up an object

Node
A point on a graph

Octahedron
A solid with eight faces

Parabola
The shape given by a $y = x^2$ graph

Percentage
A fraction out of one hundred

Photon
A massless particle that forms the electromagnetic spectrum, including visible light

Probability
The likelihood of an event occurring

Projectile
An unpowered object that has been launched, thrown or fired

Radius
The distance from the centre of a circle to its edge

Revolution
A complete 360° rotation

rpm
An abbreviation for revolutions per minute

Sequence
A list of numbers, linked by a mathematical rule

Series
A sequence where the terms are added together

Solstice
The time when the day is longest (summer solstice) or the shortest (winter solstice)

Sphere
A ball

Surface Area
The combined area of the faces of a three-dimensional shape

Term
An item in a list, sequence or series

Tetrahedron
A triangular-based pyramid

Vacuum
An area that contains very little or no matter; empty space

Volume
The amount of space a three-dimensional shape occupies

Wavelength
The distance from the peak of a wave to the next peak

Acknowledgements

This book went through several iterations before coalescing as the work you have in your hands. For getting the project off the ground, huge thanks to Jo Stansall. The finished product is testimony to the careful guidance, encouragement and sheer elbow-grease of Gabriella Nemeth. Cheers Gabby!

This book would be far worse off without the excellent cartoons and diagrams. Neil Williams deserves much credit for taking my very amateur, imprecise scratchings and turning them – after endless tweaking from me – into works of art. Thanks a lot, Neil.

A huge quantity of unconditional support came from my dogs during the long hours at the keyboard. Thank you to Bonsai and Marble for their excess body heat and knowing when their human needs a walk.

Finally, infinite gratitude to my wife, Morag, *sine qua non.*